Nanofluids for Heat Exchangers

Hafiz Muhammad Ali · Ali Hassan · Abdul Wahab

Nanofluids for Heat Exchangers

Springer

Hafiz Muhammad Ali
Mechanical Engineering Department
King Fahd University of Petroleum
and Minerals
Dhahran, Saudi Arabia

Interdisciplinary Research Center for
Renewable Energy and Power Systems
(IRC-REPS)
King Fahd University of Petroleum
and Minerals
Dhahran, Saudi Arabia

Abdul Wahab
Department of Engineering
Durham University
Durham, UK

Ali Hassan
School of Mechanical, Medical and Process
Engineering
Queensland University of Technology
Brisbane, QLD, Australia

ISBN 978-981-19-3226-7 ISBN 978-981-19-3227-4 (eBook)
https://doi.org/10.1007/978-981-19-3227-4

This Springer imprint is published by the registered company Springer Nature Singapore Pte Ltd.
The registered company address is: 152 Beach Road, #21-01/04 Gateway East, Singapore 189721, Singapore

Hafiz Muhammad Ali dedicates this book to his beloved father Muhammad Arshad Janjua

Acknowledgment

Hafiz Muhammad Ali thankfully acknowledges the support of King Fahd University of Petroleum and Minerals, Dhahran, Saudi Arabia.

Contents

Nomenclature

NDG	*Nitrogen doped Graphene*
DI	*Deionized*
CTAB	*Cetyl Trimethyl Ammonium Bromide*
OHTC	*Overall Heat Transfer Coefficient*
HTCR	*Heat Transfer Coefficient Ratio*
SLS	*Sodium Lauryl Sulfate*
HTC	*Heat Transfer Coefficient*
PPC	*Parallel Plate channel*
PEC	*Performance Evaluation Criteria*
AR	*Aspect Ratio*
GA	*Gum Arabic*
φ	*Volume concentration*
P	*Helical Pitch*
μ_w	*Fluid viscosity at tube wall*
$\frac{y_o}{y}$	*Overlapped Twist Ratio*
ρ_{bf}	*Base fluid density*
m_{np}	*Mass of nanoparticle*
$Cp\ bf$	*Base fluid Specific Heat*
N	*Number of passes per stream*
b_c	*Mean spacing between plates*
μbf	*Base fluid viscosity*
F	*Temperature Correction factor*
d	*Tube side diameter*
C_p	*Hot and Cold fluid Specific heat*
Pe	*Peclet number*
$C_{\min}$	*Minimum Capacity Rate*
f	*Friction factor*
V	*Volumetric flow rate*
SDBS	*Sodium dodecyl benzene sulfonate*
PVP	*Polyvinyl Pyrrolidine*
HTRR	*Heat Transfer Rate Ratio*

OHTCR	*Overall Heat Transfer Coefficient Ratio*
A′ C′	*Amorphous Carbonic*
CWC	*Corrugated Wavy Channel*
TPF	*Thermal Performance Factor*
AC	*Activated Carbon*
CMC	*Carboxymethyl Cellulose*
h	*Hot fluid flow*
C'	*Percent concentration*
D	*Diameter of helical coil*
μ_b	*Fluid viscosity at bulk mean temperature*
ρ_{np}	*Nanoparticle density*
$Cp\ np$	*Nanoparticle Specific Heat*
m_{bf}	*Mass of basefluid*
u_s	*Superficial velocity*
w	*Plate width inside gasket*
k	*Thermal Conductivity*
β	*Chevron angle of PHE*
L	*Length*
μnf	*Nanofluid viscosity*
T_a	*Thermodynamic average Temperature*
fg	*Flue gas*
G	*Specific flow rate*

Chapter 1
Shell and Tube Heat Exchanger

Abstract Shell and tube heat exchanger are crucial components in any industry. In this chapter, effect on thermal performance of shell and tube heat exchanger by using different nanofluids in terms of overall heat transfer coefficient, Nusselt number, pressure drop, NTU, friction factor, entropy generation, energy, and exergy efficiency has been reviewed under different conditions such as volume concentration of nanofluids, mass flow rate, Reynold number, inlet temperature of working fluid and geometry of nanoparticles. Energy, exergy and entropy generation relations used in different literatures are mentioned in this chapter.

1.1 Introduction

For chemical and industrial processes, heat exchangers especially shell and tube heat exchanger are one of the important and mostly used components that can be found in refineries, power plants, pharmaceuticals, air conditioning, manufacturing industries, process industries and food industries. So, it is necessary to enhance heat transfer rate to achieve economic and environmental advantages. Heat transfer in shell and tube heat exchanger can be done either by enhancing area of shell and tube by introducing different geometries on tube side like fins, twisted tapes, corrugated tubes and wire coils and other way is to use such heat transferring fluid with superior heat transfer properties compared to conventional fluids.

Hosseini et al. [1] compared performance of micro finned, corrugated and smooth tube in shell and tube heat exchanger. Micro finned tube outsmarted the other two tubes in terms of Nu number. Result also showed that corrugation tube showed negligible increment in Nu number, but pressure drop was similar to that of smooth tube. Dizaji et al. [2] employed corrugated tube and corrugated shell to analyze exergy loss and NTU. Exergy loss and NTU were increased from 17 to 81% and 34 to 60% when corrugated shell and tube were used. Marzouk et al. [3] enhanced effectiveness (185–224%), NTU (132–149%), exergy efficiency (130–210%) and overall heat transfer coefficient (210–280%) by introducing wired nails with circular rods on the tube of shell and tube heat exchanger. NTU and friction factor was raised

H. M. Ali et al., *Nanofluids for Heat Exchangers*,
https://doi.org/10.1007/978-981-19-3227-4_1

up to 2.64 times and 2.74 times when helical wires were used inside the tube of heat exchanger [4]. Zhang and Liu et al. [5] numerically generated vortex effect inside the tube using triple and quadruple twisted tape which helped in increasing Nu number by 171 and 182% with increment of 4.06–7.02 times the friction factor.

Wang et al. [6] modified design of shell and tube heat exchanger by employing sealers between baffle and shell. This modification helped in mitigating baffle-shell gap and circular leakage flow which caused increment in heat transfer coefficient on shell side (18.2–25.5%), pumping power (44.6–48.8%), overall heat transfer coefficient (15.6–19.1%) and exergy coefficient (12.9–14.1%). Miansari et al. [7] simulated performance of helical shell and tube heat exchanger with and without fins. ST heat exchanger showed better performance with circular fins compared to without finned heat exchanger. Result showed that efficiency was reduced by raising the hot fluid velocity while efficiency was enhanced by raising cold fluid velocity. Heat transfer was also improved by flowing hot fluid through coils while cold fluid through shell, respectively. Arani and Moradi et al. [8] numerically optimized performance of shell and tube heat exchanger by introducing disk baffle, combined segmental disk baffle and circular and triangular ribbed tubes. Study showed that pressure drop was reduced by employing disk baffle and combined segmental disk baffle while heat transfer rate was improved by triangular ribbed tube due to enhanced heat transfer area and turbulence effect. Among these, disk baffle with triangular ribbed tube showed best performance. Pandiyarajan et al. [9] used finned shell and tube heat exchanger to recover heat from exhaust of diesel engine using castor oil as heat transfer fluid on tube side with hot exhaust gas flowing through shell side. It was found that heat extraction rate was reduced with time due to increment in castor oil temperature. About 3.6 kW of heat was recovered from the exhaust gas using finned shell and tube heat exchanger. Movassag et al. [10] used helical baffle to replace conventional and segmental baffles. Helical baffles helped in reducing pressure drop and fouling. In another study by Gao et al. [11], it was suggested to use helical baffle with small helix angle according to the analyses based on 2nd law of thermodynamics while baffle with greater helical angle showed better performance in terms of energy analysis. Stehlik et al. [12] suggested that space between baffles, helix angle and arrangement of baffles are the important parameters for optimizing performance of shell and tube heat exchanger with helical baffles.

1.2 Performance of Nanofluid in Shell and Tube Heat Exchanger

Shell and tube heat exchanger mainly consists of shell with number of tubes present inside the shell. Working fluid flowing inside heat exchanger is water and other base fluids like oils and ethylene glycol. A detailed overview of recent research conducted on shell and tube heat exchanger using nanofluid is given in Fig. 1.1.

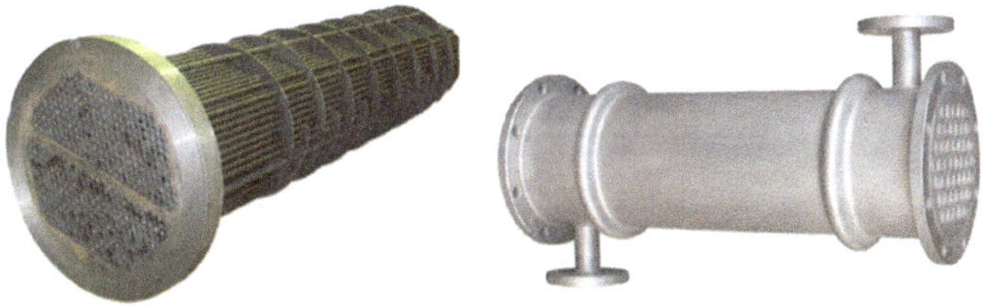

Fig. 1.1 Shell and tube heat exchanger [13]

Farajollahi et al. [14] used two different nanofluids and compared their results. Results showed 56% increment in convective heat transfer coefficient which was achieved at 0.3 and 0.5% volume concentration for TiO_2/H_2O and γ-Al_2O_3/H_2O, respectively. TiO_2/H_2O nanofluid showed better heat transfer properties than γ-Al_2O_3/H_2O at lower concentration (<0.3 vol. %) while above 0.3% concentration γ-Al_2O_3/H_2O showed better heat transfer properties. Albadr et al. [15] studied forced convective heat transfer in horizontal shell and tube heat exchanger using Al_2O_3/H_2O nanofluid. Friction factor was decreased by enhancing Reynold number at lower volume concentration. Nusselt number and overall heat transfer coefficient was enhanced by increasing Peclet number. But heat transfer coefficient of water was raised up to optimum value of Peclet number and then started to decrease. Nusselt number and heat transfer coefficient was 62.6 and 57% greater than base fluid at 2% volume concentration and 0.0125 L/s flow rate, respectively. Ghozatloo et al. [16] used graphene/water nanofluid to study heat convective heat transfer in shell and tube heat exchanger. Nanofluid was more efficient at higher temperature than at lower temperature. Results showed that when temperature of fluid was enhanced from 25 to 38 °C heat transfer was increased by 13.1% at 0.1% concentration. Thermal conductivity was enhanced by increasing weight concentration up to optimum value. Duangthongsuk and Wongwises et al. [17] enhanced heat transfer in horizontal double tube counter flow heat exchanger using TiO_2/H_2O nanofluid. Heat transfer coefficient was enhanced by reducing nanofluid temperature and increasing flow rate of hot water. Hot water temperature had no noticeable effect on heat transfer coefficient. Results also showed that pressure drop of nanofluid was almost equal to that of water. Heat transfer coefficient of base fluid (water) was less than nanofluid by 6–11%.

Huminic and Huminic et al. [18] numerically showed that outlet water temperature was raised by increasing volume concentration of nanoparticle. Heat transfer coefficient achieved at 2% concentration was more than 3% concentration because at higher concentration effect of viscosity was more dominant than thermal conductivity effect. Leong et al. [19] used shell and tube heat exchanger in biomass heating plant along with nanofluid (Cu/EG) as a coolant. No heat transfer was achieved in turbulent flow when water-based copper nanofluid was used. Heat transfer was enhanced by enhancing flow rate of flue gas, while nanofluid flow rate had a minor effect on heat transfer. 15.97 and 0.71% enhancement in heat transfer was noticed at 42 kg/s flue gas flow rate and 230 kg/s coolant mass flow rate by using Cu/EG

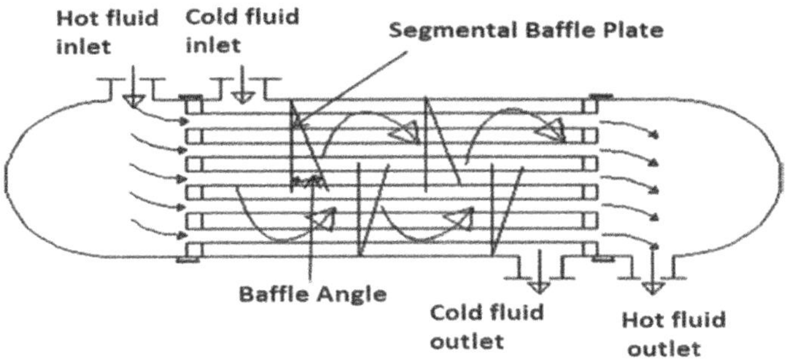

Fig. 1.2 Baffle angle in shell and tube heat exchanger [20]

nanofluid, respectively. Elias et al. [20] investigated effect of different nanoparticles shape (platelets, blades, bricks and cylindrical) on shell and tube HE performance at various baffle angle (20°, 30°, 40°, 50°). Maximum heat transfer was achieved by cylindrical shaped boehmite alumina (γ-AlOOH/EG-H_2O) at 20° baffle angle. Also, entropy generation reduction was higher at 20° baffle angle and cylindrical shaped nanoparticles. Heat transfer rate was enhanced by 22.3% for cylindrical nanoparticle at 1% concentration and 20° baffle angle in comparison with 40° baffle angle. Shell and tube heat exchanger schematic with baffle angle is given in Fig. 1.2.

Akhtari et al. [21] compared performance of double pipe and shell & tube heat exchanger using α-Al_2O_3/H_2O nanofluid. Heat transfer coefficient was increased by increasing hot and cold fluid flow rate and nanofluid temperature because chaotic movement of nanoparticles in water was enhanced which helped in increasing heat transfer rate. Shell and tube heat exchanger had 26% higher heat transfer rate than double pipe heat exchanger. Godson et al. [22] used shell and tube heat exchanger at the outlet of flat plate solar collector. Nanofluid was used in flat plate collector and then in shell and tube heat exchanger. Results showed that heat transfer coefficient was increased at higher volume fraction and flow rate because due to use of nanofluid, thermal conductivity was enhanced with delay in boundary layer formation. 12.4, 10.87 and 9.2% HTC was raised at 0.04, 0.03 and 0.01% volume concentration, respectively. Bahiraei et al. [23] numerically analyzed effect of baffle overlapping and helix angle on performance of shell and tube heat exchanger using two-phase model and Al_2O_3/H_2O nanofluid. Enhancement in HTC was greater at low concentration than at higher concentration due to non-linear relation of concentration and thermal conductivity. Heat transfer rate and pressure drop was directly proportional to baffle overlapping, volume concentration and inversely related to helix angle. Helical Baffle given in Fig. 1.3.

Lotfi et al. [24] synthesized MWCNT by using catalytic chemical vapor deposition method. MWCNT was made hydrophilic by adding COOH functional groups. 6.67% enhancement in heat transfer coefficient was achieved in comparison with water at 0.015% concentration. Shahrul et al. [25] studied that no surfactant was required to prepare stable SiO_2/H_2O and Al_2O_3/H_2O nanofluid while PVP was used for preparing stable ZnO/H_2O and Fe_3O_4 was unstable for all surfactants used (PVP,

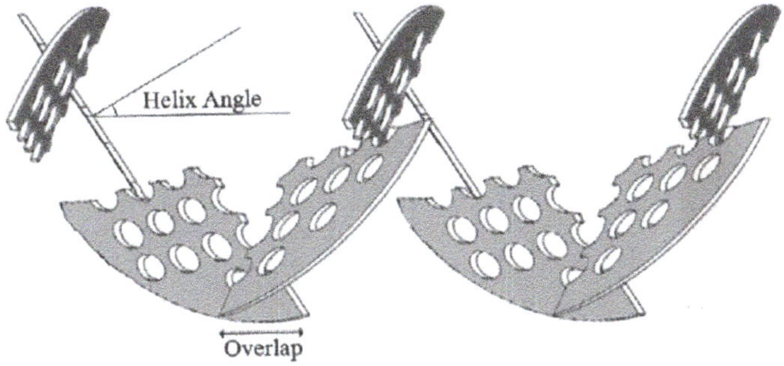

Fig. 1.3 Helical baffle schematic [23]

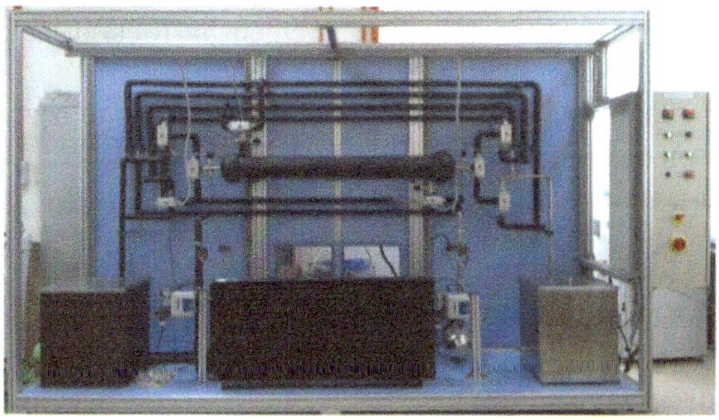

Fig. 1.4 Experimental rig of shell and tube heat exchanger [25]

CTAB SDS). OHTC was increased by 12, 26 and 35% for SiO_2/H_2O (0.5% concentration), Al_2O_3/H_2O (0.5% concentration) and ZnO/H_2O (0.3% concentration) with PVP as surfactant, respectively. While maximum increment of 50% in HTC was observed for ZnO/H_2O at 0.3% (Fig. 1.4).

Kumar and Sonawane et al. [26] showed that Fe_2O_3/H_2O had higher thermal conductivity than Fe_2O_3/EG due to low viscosity of Fe_2O_3/H_2O which allowed rapid movement of particles and enhanced Brownian motion. 20 and 29% enhancement in convective heat transfer coefficient and Nu number was achieved for Fe_2O_3/H_2O at 3 LPM and 0.08% concentration, respectively, while 13 and 14% enhancement in convective heat transfer coefficient and Nu number was achieved for Fe_2O_3/EG at 3 LPM, 0.08% concentration. No increment in pressure drop was noted in laminar flow condition. Aghabozorg et al. [27] experimentally studied magnetic hybrid nanofluid (Fe_2O_3-CNT/H_2O) in horizontal shell and tube heat exchanger under different voltages (80 V, 120 V, 150 V). Improvement in heat transfer was greater in turbulent flow than laminar flow. Also, increasing voltage increased heat transfer rate due to reduced shear stress and thermal boundary layer thickness. Heat transfer coefficient was enhanced by 37.50 and 34.02 at 0.2% concentration under turbulent and laminar flow, respectively. Barzegarian et al. [28] prepared stable gamma-Al_2O_3/H_2O nanofluid with help of SDBS surfactant. Stable nanofluid helped in increment of 18.9% in thermal performance factor at 0.3% volume fraction. Only drawback of

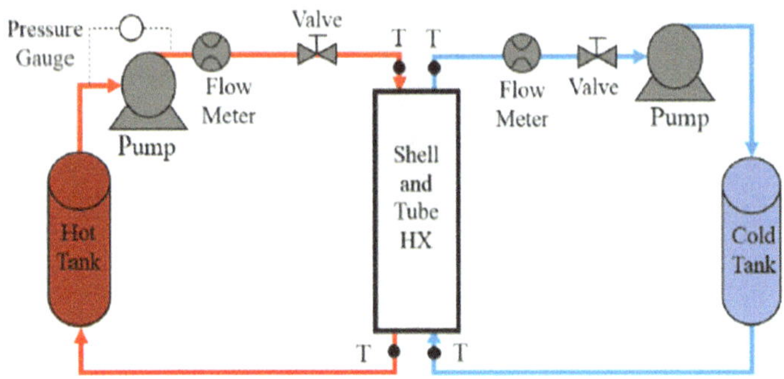

Fig. 1.5 Schematic diagram of shell and tube heat exchanger setup [30]

using nanofluid was 22% increment in friction factor which raised pressure drop resulted in increment in pumping power.

Shahrul et al. [29] found that effectiveness of heat exchanger was directly dependent on shell side (hot water) flow rate and inversely related to tube side (nanofluid) flow rate. Among studied nanofluid, ZnO showed maximum energy effectiveness followed by CuO/H_2O, Fe_3O_4/H_2O, TiO_2/H_2O, Al_2O_3/H_2O because of maximum thermal conductivity as compared to given nanofluids. On the other hand, Al_2O_3/H_2O and CuO/H_2O had maximum and minimum heat transfer coefficient enhancement, respectively. Overall heat transfer coefficient was raised to 22 and 18.35 W/m^2K when flow rate of hot side (hot water) and cold side (nanofluid) was enhanced to 70 kg/min in presence of Al_2O_3/H_2O, respectively. Esfahani et al. [30] found that graphene oxide nanofluid showed now-Newtonian behavior at 0.1% weight concentration. Also, nanoparticle size was increased from 650 to 1220 nm when concentration was enhanced from 0.01 to 0.1% concentration. Thermal conductivity was enhanced to 20% at 0.1% concentration. Moreover, exergy loss was reduced due to decrement in inlet fluid temperature and increment in flow rate of hot fluid, respectively (Fig. 1.5).

Nallusamy et al. [31] showed that shell and tube heat exchanger was more efficient than counter and parallel flow heat exchanger. OHTCR obtained was 1.202, 1.151 and 1.148 for shell and tube, parallel and counter flow heat exchanger. OHTCR was raised by raising flow rate of nanofluid. Bahiraei et al. [32] predicted effect of Reynold number and volume fraction on Nu number and frictional factor. Friction factor was independent of volume fraction while it was enhanced by reducing Reynold number. Moreover, Nu number was enhanced by 20% when particle concentration was raised from 0 to 5% by volume. Khoddamrezaee et al. [33] showed that shear stress in external flow was enhanced due to increment in viscosity of nanofluid (Al_2O_3/H_2O) by raising the concentration. Nallusamy and Prabu et al. [34] analyzed Al_2O_3/H_2O nanofluid in parallel and counter flow heat exchanger. Average Nusselt number was raised by 10, 6.5 and 10.5% for parallel, counter flow and shell and tube heat exchanger. LMTD was increased by 15.5, 14.5 and 12.5% for parallel, counter and shell and tube heat exchanger, respectively.

Artificial Neural Network was developed to predict the performance of nanofluid (γ-Al_2O_3/H_2O, TiO_2/H_2O) in shell and tube heat exchanger. ANN model was consisted of two hidden layers with 4 neurons in each layer. Model performance was compared with experimental results in terms of Nusselt number and pressure drop. Predicted model showed variation of 9 and 9.5% variation from experimental results for Nusselt number and pressure drop [35]. Rostami et al. [36] enhanced mean effectiveness by 100 and 85% using fin blades on tubes of shell and tube heat exchanger with SWCNT and graphene quantum dot nanofluid, respectively. Kumar and Arasu et al. [37] used TiO_2–Ag/H_2O nanofluid to enhance performance of shell and tube heat exchanger. Tube side heat transfer coefficient and overall heat transfer coefficient was enhanced by increasing shell inlet temperature and flow rate. Viscosity of nanocomposite was found to be high compared to water which caused decrement in Re number and increased in Pr number and pumping power. Fares et al. [38] used graphene water nanofluid as hot stream on tube side to analyze its thermal performance. Overall heat transfer coefficient, hot side and cold side thermal efficiency was raised by raising concentration of graphene nanofluid. While thermal efficiency of tube side was reduced by raising the flow rate of nanofluid, maximum enhancement in mean thermal efficiency observed was 13.7% at 0.2 wt.% and 2.16 l/min.

Sajjad et al. numerically analyzed thermal hydraulic performance of ZrO_2/H_2O nanofluid using Gnielinski correlation (given below) on tube side and Bell Delaware method (explained in introduction) employed on shell side. Numerical results revealed that overall heat transfer coefficient was positively affected by enhancing tube side Re number, volume concentration and mass flow rate on shell side. Also, triangular tube configuration had superior performance than square tube in terms of heat transfer coefficient.

$$\mathrm{Nu}_t = 0.012\left(\mathrm{Re}_t^{0.87} - 280\right)\mathrm{Pr}_t^{0.4} \tag{1.1}$$

Arulkumar and Mathanraj et al. [39] used titanium nitride water based nanofluid for achieving better thermal performance for shell and tube heat exchanger. LMTD, efficiency and effectiveness of heat exchanger was raised by raising the weight concentration of nanofluid. Karimi et al. [40] analyzed performance of shell and tube heat exchanger by employing SiC/H_2O nanofluid on tube side while hot water on the shell side. It was noted that overall heat transfer coefficient, heat transfer coefficient inside the tube and Nu number were enhanced by raising flow rate of nanofluid inside the tube. Sridhar et al. [41] compared thermal performance of SnO_2/H_2O and Ag/H_2O nanofluid. Ag/H_2O found to be superior in terms of Nu number, convective heat transfer coefficient along with pressure drop and friction factor due to high thermal conductivity, viscosity and density compared to SnO_2/H_2O nanofluid. Goldanlou et al. [42] numerically investigated the optimum conditions for shell and tube heat exchanger equipped with blade turbulators in terms of Re, volume concentration and blade rod diameter on the basis of performance evaluation criteria (PEC). It was found that PEC was enhanced by increasing Re number and concentration of hybrid nanofluid (Fe_3O_4–CNT/H_2O) while PEC first increased with blade rod diameter and then started to decrease.

$$\text{PEC} = \left(\frac{\text{Nu}_{av}}{\text{Nu}_{av,0}}\right)\left(\frac{f}{f_0}\right)^{-\frac{1}{3}} \tag{1.2}$$

where

$$f = \left(\frac{2}{\frac{L}{D_r}}\right)\left(\frac{\Delta P}{\rho_{nf} u_m^2}\right) \tag{1.3}$$

Hu et al. [43] employed CFD to numerically investigate the performance of MWCNT/H_2O nanofluid in shell and tube heat exchanger using $k-\varepsilon$ model (given below). Numerical results showed that along with increment in heat transfer rate, pressure drop across the heat exchanger was raised. Using 2% concentration of MWCNT/H_2O caused an increment of 1.6% pressure drop at 1.5 kg/s flow rate.

$$\frac{\partial}{\partial x_i}(\rho k u_i) = \frac{\partial}{\partial x_j}\left[\left(\mu + \frac{\mu_t}{\sigma_k}\right)\left(\frac{\partial k}{\partial x_j}\right)\right] + G_k + G_b - \rho\varepsilon + S_k \tag{1.4}$$

$$\begin{aligned}\frac{\partial}{\partial x_i}(\rho \varepsilon u_i) &= \frac{\partial}{\partial x_j}\left[\left(\mu + \frac{\mu_t}{\sigma_\varepsilon}\right)\left(\frac{\partial \varepsilon}{\partial x_j}\right)\right] \\ &+ C_{1\varepsilon}\left(\frac{\varepsilon}{k}\right)(G_k + C_{3\varepsilon}G_b) - C_{2\varepsilon}\rho\left(\frac{\varepsilon^2}{k}\right) + S_\varepsilon\end{aligned} \tag{1.5}$$

$$S_k = \tau_{ij}^R \frac{\partial u_i}{\partial x_j} - \rho\varepsilon + \mu_t P_B \tag{1.6}$$

$$S_\varepsilon = C_{1\varepsilon}\frac{\varepsilon}{k}\left(f_1 \tau_{ij}^R \frac{\partial u_i}{\partial x_j} + \mu_t C_B P_B\right) - C_{2\varepsilon} f_2 \frac{\rho\varepsilon^2}{k} \tag{1.7}$$

$$P_B = -\frac{g_i}{\sigma_B}\frac{1}{\rho}\frac{\partial \rho}{\partial x_i} \tag{1.8}$$

$$\tau_{ij}^R = \mu\left(\frac{\partial u_i}{\partial x_j} + \frac{\partial u_j}{\partial x_i} - \frac{2}{3}\delta_{ij}\frac{\partial u_k}{\partial x_k}\right) - \frac{2}{3}\delta_{ij}\rho k \tag{1.9}$$

$$f_1 = 1 + \left(\frac{0.05}{f_\mu}\right)^3 \tag{1.10}$$

$$f_2 = \left(1 - \exp\left(-R_T^2\right)\right) \tag{1.11}$$

$$f_\mu = \left(1 - e^{-0.0165 R_y}\right)\left(1 + \left(\frac{20.5}{R_T}\right)\right) \tag{1.12}$$

where turbulent viscosity is:

$$\mu_t = \rho C_\mu \frac{k^2}{\varepsilon} \tag{1.13}$$

Model constant have the following values [44].

$$C_{1\varepsilon} = 1.44;\ C_{2\varepsilon} = 1.92;\ C_\mu = 0.09;\ \sigma_k = 1;\ \sigma_\varepsilon = 1.3$$

Singh and Sarkar et al. [45] compared the performance of hybrid nanofluid in shell and tube condenser of nuclear power plant. Among studied nanofluid, Al_2O_3-MWCNT showed maximum reduction in flow rate with least pumping power followed by Al_2O_3–TiO_2, Al_2O_3–Cu, and Al_2O_3–Ag, respectively. While, Al_2O_3_Ag showed maximum exergy efficiency of 29.97% and Al_2O_3-MWCNT showed minimum exergy efficiency of 29.81%. For turbulent flow, Xuan and Li correlation was used for thermal analysis.

$$\mathrm{Nu_{hnf}} = 0.0059\left(1 + 7.6286\emptyset^{0.6886}\mathrm{Pe_{hnf}^{0.001}}\right)\mathrm{Re_{hnf}^{0.9238}}\mathrm{Pr_{hnf}^{0.4}} \tag{1.14}$$

1.3 Thermal Performance of Shell and Tube Heat Exchanger

1.3.1 Energy Analysis

Thermal performance of shell and tube heat exchanger is evaluated on the basis of overall convective heat transfer coefficient, log mean temperature difference (LMTD), heat transfer between hot and cold streams and thermal efficiency of shell and tube side [38].

Overall Heat Transfer coefficient can be evaluated using the following relation:

$$U = \frac{Q}{A \times \mathrm{LMTD}} \tag{1.15}$$

If correction factor for LMTD is considered then

$$U = \frac{Q}{A \times \mathrm{LMTD} \times F} \tag{1.16}$$

where F is the correction factor for LMTD [1]:

$$F = \frac{\sqrt{R^2+1}\ln\left(\frac{1-P}{1-PR}\right)}{(R-1)\ln\left[\frac{2-P\left(P+1-\sqrt{R^2+1}\right)}{2-P\left(R+1+\sqrt{R^2+1}\right)}\right]} \tag{1.17}$$

$$R = \frac{T_{h1} - T_{h2}}{T_{c2} - T_{c1}} \tag{1.18}$$

$$P = \frac{T_{c2} - T_{c1}}{T_{h1} - T_{c1}} \tag{1.19}$$

Q and LMTD are heat transfer between the fluids and log mean temperature difference.

$$Q = V\left(\rho C_p\right)_{nf}(T_{h1} - T_{h2}) \tag{1.20}$$

and LMTD is:

$$\text{LMTD} = \frac{(T_{h1} - T_{c2}) - (T_{h2} - T_{c1})}{\ln\left(\frac{T_{h1} - T_{c2}}{T_{h2} - T_{c1}}\right)} \tag{1.21}$$

T_{h1}, T_{h2}, T_{c1}, T_{c2} are the inlet and outlet temperature of hot stream (tube side) and cold stream (shell side).

Thermal efficiency of the hot side and cold side is defined below:

$$\eta_{\text{hot}} = \frac{T_{h1} - T_{h2}}{(T_{h1} - T_{c1})} \tag{1.22}$$

$$\eta_{\text{cold}} = \frac{T_{c2} - T_{c1}}{(T_{h1} - T_{c1})} \tag{1.23}$$

Mean thermal efficiency of the heat exchanger is:

$$\eta_{\text{mean}} = \frac{\eta_{\text{hot}} + \eta_{\text{cold}}}{2} \tag{1.24}$$

1.3.2 Exergy Analysis

Exergy analysis of shell and tube heat exchanger can be done using following relation [30]:

$$E = E_{\text{h}} + E_{\text{c}} \tag{1.25}$$

E_{h} and E_{c} are the exergy change of hot and cold fluid. E is the exergy loss of the open system.

$$E_{\mathrm{h}} = T_{\mathrm{e}}\left[m_{\mathrm{h}}\left(S_{\mathrm{h,\,out}} - S_{\mathrm{h,\,in}}\right)\right] \text{ OR } E_{\mathrm{h}} = T_{\mathrm{e}}\left[C_{\mathrm{h}} \ln\left(\frac{T_{\mathrm{h,\,out}}}{T_{\mathrm{h,\,in}}}\right)\right] \tag{1.26}$$

$$E_{\mathrm{c}} = T_{\mathrm{e}}\left[m_{\mathrm{c}}\left(S_{\mathrm{c,out}} - S_{\mathrm{c,in}}\right)\right] \text{ OR } E_{\mathrm{c}} = T_{\mathrm{e}}\left[C_{\mathrm{c}} \ln\left(\frac{T_{\mathrm{c,out}}}{T_{\mathrm{c,in}}}\right)\right] \tag{1.27}$$

Combining above two equations:

$$E = T_{\mathrm{e}}\left[C_{\mathrm{h}} \ln\left(\frac{T_{\mathrm{h,out}}}{T_{\mathrm{h,in}}}\right) + C_{\mathrm{c}} \ln\left(\frac{T_{\mathrm{c,out}}}{T_{\mathrm{c,in}}}\right)\right] \tag{1.28}$$

Dimensionless exergy (2nd law efficiency) can be evaluated using following relation [3]:

$$e = \frac{E}{m_{\mathrm{h}} \Delta \Psi_{\mathrm{h}}} \tag{1.29}$$

$$\Delta\Psi_{\mathrm{h}} = (h_{\mathrm{out}} - h_{\mathrm{in}}) - T_{\mathrm{e}}(S_{\mathrm{out}} - S_{\mathrm{in}}) + \frac{u_{\mathrm{out}}^2 - u_{\mathrm{in}}^2}{2} + g(Z_{\mathrm{out}} - Z_{\mathrm{in}}) \tag{1.30}$$

Depending on the working conditions, potential and kinetic energy can be neglected.

$$\Delta\Psi_{\mathrm{h}} = (h_{\mathrm{out}} - h_{\mathrm{in}}) - T_{\mathrm{e}}(S_{\mathrm{out}} - S_{\mathrm{in}}) \tag{1.31}$$

where

$$h_{\mathrm{out}} - h_{\mathrm{in}} = C \Delta T_{\mathrm{h}} \tag{1.32}$$

$$e = \frac{E}{m_{\mathrm{h}} C_{\mathrm{p}}\left[\Delta T_{\mathrm{h}} - T_{\mathrm{e}}\left(\ln\left(\frac{T_{\mathrm{h,\,out}}}{T_{\mathrm{h,\,in}}}\right)\right)\right]} \tag{1.33}$$

C_p is the specific heat of hot fluid, m_h is the mass flow rate of hot fluid and T_e is ambient temperature.

1.3.3 Entropy Generation Analysis

In any heat transfer process, there is always entropy generation due to irreversibility of the system. Generally, in shell and tube heat exchanger there are two types of entropy generations/irreversibility:

i. Heat conduction in finite temperature difference [46]
ii. Other is finite pressure drop due to flow friction [47].

Each of them can be evaluated using following relation:

$$\dot{S}_{\text{gen},\Delta T} = \int_{\text{in}}^{\text{out}} \left(\frac{q_{\text{m}} C_{\text{p}} \text{d}T}{T}\right)_{t,s} = (q_{\text{m}} C_{\text{p}})_{\text{t, s}} \ln\left(\frac{T_{\text{t,out}}}{T_{\text{t,in}}}\right) + (q_{\text{m}} C_{\text{p}})_{\text{s}} \ln\left(\frac{T_{\text{s, out}}}{T_{\text{s, in}}}\right) \quad (1.34)$$

T_s shell side temperature, T_t tube side temperature, q_m mass flow rate.

$$\dot{S}_{\text{gen},\Delta P} = \left(q_{\text{m}} \frac{\Delta P}{\rho} \frac{\ln\left(\frac{T_{\text{out}}}{T_{\text{in}}}\right)}{T_{\text{out}} - T_{\text{in}}}\right)_{\text{t, s}} = q_{\text{m, t}} \frac{\Delta P_{\text{t}}}{\rho_{\text{t}}} \frac{\ln\left(\frac{T_{\text{t, out}}}{T_{\text{t, in}}}\right)}{T_{\text{t, out}} - T_{\text{t, in}}} + q_{\text{m, s}} \frac{\Delta P_{\text{s}}}{\rho_{\text{s}}} \frac{\ln\left(\frac{T_{\text{s, out}}}{T_{\text{s, in}}}\right)}{T_{\text{s, out}} - T_{\text{s, in}}} \quad (1.35)$$

Total entropy generation will be:

$$\dot{S}_{\text{gen}} = \dot{S}_{\text{gen},\Delta P} + \dot{S}_{\text{gen},\Delta T} \quad (1.36)$$

1.4 Discussion

From Fig. 1.6, Nusselt number increases by increasing Peclet number. Increase in Nusselt number resulted an enhancement in Brownian motion and reduction in

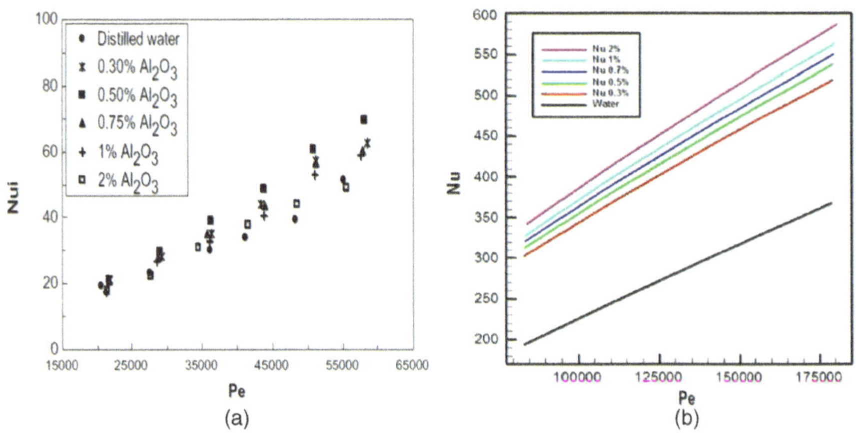

Fig. 1.6 **a** Effect of Peclet Number on Nusselt Number for Al_2O_3/H_2O at different concentrations [14]. **b** Effect of Peclet Number on Nusselt Number for Al_2O_3/H_2O at different concentrations [15]

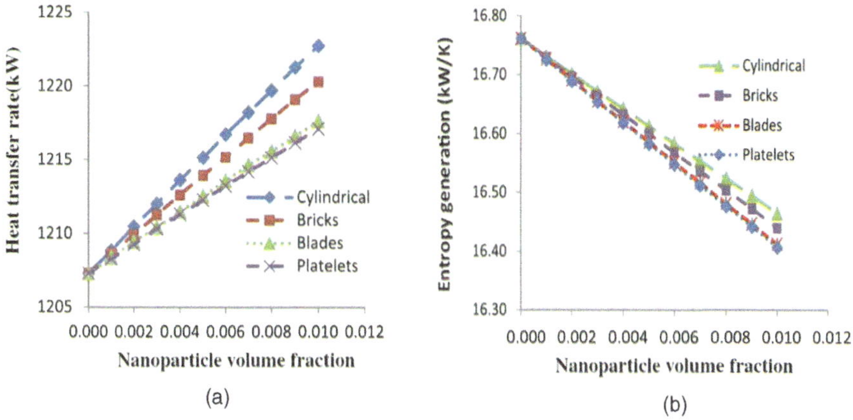

Fig. 1.7 **a** Effect of volume fraction and shape of nanoparticle on heat transfer rate. **b** Effect of volume fraction and shape of nanoparticle on entropy generation [20]

thermal boundary layer thickness which enhances the convective heat transfer coefficient. From figure, it was concluded that Nusselt number was increased by increasing volume concentration of nanoparticles because thermal conductivity and viscosity of nanofluids increase by increasing the concentration up to optimum concentration which resulted in increment in heat transfer rate. Above optimum concentration viscosity and stability effect of nanofluid dominates thermal conductivity.

Heat transfer rate and entropy generation depends on nanoparticle volume fraction and shape of nanoparticles. Heat transfer rate increases, and entropy generation decreases by enhancing the volume fraction. Figure 1.7 showed effect of different shapes on entropy generation and heat transfer rate. Cylindrical shape nanoparticle had maximum heat transfer rate as compared to other nanoparticles shape (brick, blades and platelets) because cylindrical shape had highest thermal conductivity enhancement coefficient followed by brick, blade and platelets shaped nanoparticles. Entropy generation for cylindrical shape was slightly greater than other shapes.

Table 1.1 Summary of literature on the usage of nanofluids in shell and tube heat exchanger

References	Heat exchanger	Nanofluid	Nanoparticle size	Surfactant	Concentration	Flow	Outcomes
[14]	Shell and tube	γ-Al_2O_3/H_2O, TiO_2/H_2O	25 and 10 nm	Not used	0.3–2 and 0.15–0.75 vol.%	Turbulent	56% enhancement in convective heat transfer coefficient was observed for both nanofluids
[15]	Horizontal shell and tube	Al_2O_3/H_2O	30 nm	–	0.3–2 vol.%	Turbulent	Heat transfer coefficient was 57% greater than base fluid at 2% volume concentration and 0.0125 L/s flow rate
[16]	Shell and tube exchanger	Graphene/H_2O	–	–	0.05–0.1 wt.%	Laminar	Convective heat transfer was enhanced by 13.1% at 0.1% concentration
[17]	Horizontal double tube counter flow	TiO_2/H_2O	21 nm	CTAB	0.2 vol.%	Turbulent	HTC of nanofluid was more than base fluid (water) by 6–11%
[18]	Double tube helical	CuO, TiO_2/H_2O	24 nm	–	0.5–3 vol.%	Laminar	Heat transfer coefficient achieved at 2% concentration was more than 3% concentration
[19]	Shell and tube	Cu/EG			0–1 vol.%	Turbulent	7.8% increment in heat transfer was obtained at 1% concentration, 26.3 kg s^{-1} and 116 kg s^{-1} flow rate of flue gas and nanofluid, respectively

(continued)

Table 1.1 (continued)

References	Heat exchanger	Nanofluid	Nanoparticle size	Surfactant	Concentration	Flow	Outcomes
[20]	Shell and tube	γ-AlOOH/EG-H_2O			0–1 vol.%		Heat transfer rate was enhanced by 22.3% for cylindrical nanoparticle at 1% concentration and 20° baffle angle in comparison with 40° baffle angle
[21]	Double pipe, shell and tube	α-Al_2O_3/H_2O	80 nm		0.2 and 0.5 vol.%	Laminar	HTC for shell and tube and double pipe HE was 21.3 and 13.2% higher for nanofluid as compared to water, respectively
[22]	Counter flow shell and tube	Ag/H_2O	54 nm	PVP	0.01–0.04 vol.%	Turbulent	Pressure drop was increased by 16.22% at 0.04% concentration and Re = 25,000 in comparison with water
[23]	Shell and tube	Al_2O_3/H_2O			1–5 vol.%		At 50° helix angle and 0.6 baffle overlapping, pressure drop increased to 38.6%, respectively
[24]	Horizontal shell and tube	MWCNT/H_2O			0.0115 wt.%		6.67% increment in heat transfer coefficient was achieved in contrast with water at 0.015% concentration

(continued)

Table 1.1 (continued)

References	Heat exchanger	Nanofluid	Nanoparticle size	Surfactant	Concentration	Flow	Outcomes
[25]	Counter flow shell and tube	Al_2O_3, SiO_2, ZnO/H_2O	13 Al_2O_3, 10–20 SiO_2, 100 ZnO	PVP	0.3 vol.% ZnO, 0.5 vol.% Al_2O_3, SiO_2	Laminar	ZnO was best performer among other nanofluids (Al_2O_3/H_2O and SiO_2/H_2O) with 50 and 34% increment in HTC and OHTC
[26]	Shell and tube	Fe_2O_3/H_2O, Fe_2O_3/EG	22 nm	Not used	0.02–0.08 vol.%	Laminar-Turbulent	20% and 29% increment in convective heat transfer coefficient and Nu number was achieved for Fe_2O_3/H_2O at 3 LPM, 0.08% concentration, respectively
[27]	Horizontal shell and tube	Fe_2O_3-CNT/H_2O	30 nm	Polyacrylic acid	0.1 and 0.2 wt.%	Laminar-Turbulent	Heat transfer coefficient was enhanced by 37.50 and 34.02% at 0.2% concentration under turbulent flow
[28]	Shell and tube with segmental baffle	γ-Al_2O_3/H_2O	15 nm	SDBS	0.03–0.3 vol.%	Laminar	About 29.8, 20.9 and 9.7% enhancement in Nu number was achieved at 0.3, 0.14 and 0.03% volume fraction, respectively

(continued)

Table 1.1 (continued)

References	Heat exchanger	Nanofluid	Nanoparticle size	Surfactant	Concentration	Flow	Outcomes
[29]	Shell and tube	ZnO, Al_2O_3, Fe_3O_4, CuO, TiO_2/H_2O	50, 20, 36, 40, 21	–	0.03 vol.%	Turbulent	Heat transfer coefficient was raised from 22 W/m^2K when flow rate of hot side was enhanced to 70 kg/min in presence of Al_2O_3/H_2O on tube side
[30]	Shell and tube	Graphene Oxide/H_2O	650–1220 nm		0.01 and 0.1 wt.%	Laminar & Turbulent	Thermal conductivity was enhanced to 20% at 0.1% concentration
[31]	Parallel, counter shell & tube	Al_2O_3/H_2O	50 nm		1 vol.%	Laminar	12, 14 and 16% increment in LMTD was observed for shell & tube, counter and parallel flow heat exchanger, respectively
[32]	Shell and tube with helical baffles	Al_2O_3/H_2O	40 nm		1–5 vol.%		Nu number was enhanced by 20% when particle concentration was raised from 0 to 5% by volume
[33]	Shell and tube	Al_2O_3/EG					Shear stress in external flow could be avoided due to large increment in heat transfer rate

(continued)

Table 1.1 (continued)

References	Heat exchanger	Nanofluid	Nanoparticle size	Surfactant	Concentration	Flow	Outcomes
[34]	Parallel and counter flow shell and tube	Al_2O_3/H_2O	50 nm	Not used	1 vol.%		LMTD was increased by 15.5, 14.5 and 12.5% for parallel, counter and shell and tube heat exchanger, respectively
[35]	Shell and tube	(γ-Al_2O_3/H_2O, TiO_2/H_2O	25 and 10 nm	Not used	0.3–2 vol.% and 0.15–0.75 vol.%		Developed ANN model showed that nanofluid had 30% higher Nu number than base fluid
[36]	Shell and tube with fin blade	SWCNT and graphene quantum dot	N-Methyl pyrrolidone for SWCNT		1–5 vol.%		Thermal properties (heat transfer rate, tube side heat transfer coefficient) were decreased by increasing concentration more than 1%
[37]	Shell and tube	TiO_2–Ag/distilled water		65 nm	0.2 vol.%		Overall heat transfer coefficient and heat transfer rate using nanofluid and water was found to be 549.36 and 489.8 W/m^2K, 1347 and 1221 W, respectively

(continued)

Table 1.1 (continued)

References	Heat exchanger	Nanofluid	Nanoparticle size	Surfactant	Concentration	Flow	Outcomes
[38]	Shell and tube	Graphene/H_2O	Gum Arabic	50–140 nm	0.01–0.2 wt.%		Overall heat transfer coefficient, thermal efficiency of hot and cold side was enhanced by 29, 24.4 and 7.3% at wt.% of 0.2%
[39]	Shell and tube	TiN_2/H_2O	NaOH		0.5–2 vol.%		45.32–51.3% increment in thermal efficiency was observed when concentration was raised from 0.5 to 2 vol.% at 1.5 kg/s flow rate
[40]	Shell and tube	SiC/H_2O	SDS	50 nm	0.25–1 vol.%		Rate of heat transfer raised by around 19% by using SiC/H_2O nanofluid
[41]	Shell and tube	SnO_2/H_2O and Ag/H_2O	CTAB and PVP	50 nm	0.05–0.15 vol.%		29 and 39% increment in thermal conductivity was observed by the addition of 0.1% volume concentration of SnO_2/H_2O and Ag/H_2O nanofluid

(continued)

Table 1.1 (continued)

References	Heat exchanger	Nanofluid	Nanoparticle size	Surfactant	Concentration	Flow	Outcomes
[42]	Shell and tube with blade shaped turbulators	Fe_3O_4-CNT/H_2O			0.1–0.9 vol.%	$6000 < Re < 8000$	Maximum PEC achieved was for Re = 9000, 0.9 and 1.35 vol.% for CNT and Fe_3O_4, and d_r = 15 mm
[43]	Shell and tube	MWCNT/H_2O			2 and 4 vol.%	Turbulent	29.5% increment in heat transfer rate was observed by using 4% volume concentration of MWCNT
[45]	Shell and tube condenser	Al_2O_3-MWCNT, Al_2O_3–Ag, Al_2O_3–Cu, Al_2O_3–TiO_2			0.1–1 vol.%	Turbulent	Al_2O_3–Ag showed maximum payback period time of 247 years followed by Al_2O_3–Cu (90 years), Al_2O_3-MWCNT (63 years) and Al_2O_3–TiO_2 (9.8 years), respectively

References

1. Hosseini, R., Hosseini-Ghaffar, A., & Soltani, M. (2007). Experimental determination of shell side heat transfer coefficient and pressure drop for an oil cooler shell-and-tube heat exchanger with three different tube bundles. *Applied Thermal Engineering, 27*(5–6), 1001–1008.
2. Dizaji, H. S., Jafarmadar, S., & Asaadi, S. (2017). Experimental exergy analysis for shell and tube heat exchanger made of corrugated shell and corrugated tube. *Experimental Thermal and Fluid Science, 81*, 475–481.
3. Marzouk, S., et al. (2020). Effect of wired nails circular–rod inserts on tube side performance of shell and tube heat exchanger: Experimental study. *Applied Thermal Engineering, 167*, 114696.
4. Akpinar, E. K. (2006). Evaluation of heat transfer and exergy loss in a concentric double pipe exchanger equipped with helical wires. *Energy Conversion and Management, 47*(18–19), 3473–3486.
5. Zhang, X., Liu, Z., & Liu, W. (2012). Numerical studies on heat transfer and flow characteristics for laminar flow in a tube with multiple regularly spaced twisted tapes. *International Journal of Thermal Sciences, 58*, 157–167.
6. Wang, S., Wen, J., & Li, Y. (2009). An experimental investigation of heat transfer enhancement for a shell-and-tube heat exchanger. *Applied Thermal Engineering, 29*(11–12), 2433–2438.
7. Miansari, M., et al. (2021). Thermal performance of a helical shell and tube heat exchanger without fin, with circular fins, and with V-shaped circular fins applying on the coil. *Journal of Thermal Analysis and Calorimetry, 143*(6), 4273–4285.
8. Arani, A. A. A., & Moradi, R. (2019). Shell and tube heat exchanger optimization using new baffle and tube configuration. *Applied Thermal Engineering, 157*, 113736.
9. Pandiyarajan, V., et al. (2011). Experimental investigation on heat recovery from diesel engine exhaust using finned shell and tube heat exchanger and thermal storage system. *Applied Energy, 88*(1), 77–87.
10. Movassag, S. Z., et al. (2013). Tube bundle replacement for segmental and helical shell and tube heat exchangers: Performance comparison and fouling investigation on the shell side. *Applied Thermal Engineering, 51*(1–2), 1162–1169.
11. Gao, B., et al. (2015). Experimental study of effects of baffle helix angle on shell-side performance of shell-and-tube heat exchangers with discontinuous helical baffles. *Experimental Thermal and Fluid Science, 68*, 48–57.
12. Stehlik, P., et al. (1994). Comparison of correction factors for shell-and-tube heat exchangers with segmental or helical baffles. *Heat Transfer Engineering, 15*(1), 55–65.
13. Yazıcıoğlu, A. G., et al. (2018). 4.2 heat exchangers. In Dincer, I. (Ed.), *Comprehensive energy systems* (pp. 40–69). Elsevier.
14. Farajollahi, B., et al. (2010). Heat transfer of nanofluids in a shell and tube heat exchanger. *International Journal of Heat and Mass Transfer, 53*(1–3), 12–17.
15. Albadr, J., Tayal, S., & Alasadi, M. (2013). Heat transfer through heat exchanger using Al_2O_3 nanofluid at different concentrations. *Case Studies in Thermal Engineering, 1*(1), 38–44.
16. Ghozatloo, A., et al. (2014). Convective heat transfer enhancement of graphene nanofluids in shell and tube heat exchanger. *Experimental Thermal and Fluid Science, 53*, 136–141.
17. Duangthongsuk, W., Wongwises, S., & Transfer, M. (2009). Heat transfer enhancement and pressure drop characteristics of TiO_2-water nanofluid in a double-tube counter flow heat exchanger. *International Journal of Heat and Mass Transfer, 52*(7–8), 2059–2067.
18. Huminic, G., Huminic, A., & Transfer, M. (2011). Heat transfer characteristics in double tube helical heat exchangers using nanofluids. *International Journal of Heat and Mass Transfer, 54*(19–20), 4280–4287.
19. Leong, K., et al. (2012). Modeling of shell and tube heat recovery exchanger operated with nanofluid based coolants. *International Journal of Heat and Mass Transfer, 55*(4), 808–816.
20. Elias, M., et al. (2014). Effect of different nanoparticle shapes on shell and tube heat exchanger using different baffle angles and operated with nanofluid. *International Journal of Heat and Mass Transfer, 70*, 289–297.

21. Akhtari, M., Haghshenasfard, M., & Talaie, M. (2013). Numerical and experimental investigation of heat transfer of α-Al2O3/water nanofluid in double pipe and shell and tube heat exchangers. *Numerical Heat Transfer, Part A: Applications, 63*(12), 941–958.
22. Godson, L., et al. (2014). Heat transfer characteristics of silver/water nanofluids in a shell and tube heat exchanger. *14*(3), 489–496.
23. Bahiraei, M., Hangi, M., & Saeedan, M. J. E. (2015). A novel application for energy efficiency improvement using nanofluid in shell and tube heat exchanger equipped with helical baffles. *93*, 2229–2240.
24. Lotfi, R., et al. (2012). Experimental study on the heat transfer enhancement of MWNT-water nanofluid in a shell and tube heat exchanger. *39*(1), 108–111.
25. Shahrul, I., et al. (2016). Experimental investigation on Al_2O_3–W, SiO_2–W and ZnO–W nanofluids and their application in a shell and tube heat exchanger. *International Journal of Heat and Mass Transfer, 97*, 547–558.
26. Kumar, N., et al. (2016). Experimental study of Fe_2O_3/water and Fe_2O_3/ethylene glycol nanofluid heat transfer enhancement in a shell and tube heat exchanger. *78*, 277–284.
27. Aghabozorg, M. H., et al. (2016). Experimental investigation of heat transfer enhancement of Fe_2O_3-CNT/water magnetic nanofluids under laminar, transient and turbulent flow inside a horizontal shell and tube heat exchanger. *72*, 182–189.
28. Barzegarian, R., et al. (2017). Thermal performance augmentation using water based Al_2O_3-gamma nanofluid in a horizontal shell and tube heat exchanger under forced circulation. *86*, 52–59.
29. Shahrul, I., et al. (2014). Effectiveness study of a shell and tube heat exchanger operated with nanofluids at different mass flow rates. *65*(7), 699–713.
30. Esfahani, M. R., & Languri, E. M. (2017) Exergy analysis of a shell-and-tube heat exchanger using graphene oxide nanofluids. *83*, 100–106.
31. Nallusamy, S. (2017). Characterization of Al_2O_3/water nanofluid through shell and tube heat exchangers over parallel and counter flow. Journal of Nano Research.
32. Bahiraei, M., Hosseinalipour, S. M., & Saeedan, M. (2015). Prediction of Nusselt number and friction factor of water-Al_2O_3 nanofluid flow in shell-and-tube heat exchanger with helical baffles. *Chemical Engineering Communications, 202*(2), 260–268.
33. Khoddamrezaee, F., Motallebzadeh, R., & Jajali, D. (2010). Simulation of (EG/Al_2O_3) nanofluid through the shell and tube heat exchanger with rectangular arrangement of tubes and constant heat flux. *10*(6), 500–505.
34. Nallusamy, S., & Manikanda Prabu, N. (2017) Heat transfer enhancement analysis of Al_2O_3-water nanofluid through parallel and counter flow in shell and tube heat exchangers. *16*(05n06), 1750020.
35. Hojjat, M. (2020). Nanofluids as coolant in a shell and tube heat exchanger: ANN modeling and multi-objective optimization. *Applied Mathematics and Computation, 365*, 124710.
36. Rostami, M. H., et al. (2020). Thermal performance investigation of SWCNT and graphene quantum dots nanofluids in a shell and tube heat exchanger by using fin blade tubes. *Heat Transfer, 49*(8), 4783–4800.
37. Kumar, D. D., & Arasu, A. V. (2021). Experimental investigation on dimensionless numbers and heat transfer in nanocompositefluid shell and tube heat exchanger. *Journal of Thermal Analysis and Calorimetry, 143*(2), 1537–1553.
38. Fares, M., Mohammad, A.-M., & Mohammed, A.-S. (2020). Heat transfer analysis of a shell and tube heat exchanger operated with graphene nanofluids. *Case Studies in Thermal Engineering, 18*, 100584.
39. Arulkumar, S., & Mathanraj, V. (2020). Heat transfer enhancement with titanium nitride nanofluid in a shell and tube heat exchanger. In *IOP Conference Series: Materials Science and Engineering*. IOP Publishing.
40. Karimi, S., et al. (2020). Experimental investigation of convective heat transfer and pressure drop of SiC/water nanofluid in a shell and tube heat exchanger. *Heat and Mass Transfer, 56*(8), 2325–2331.

41. Sridhar, S., Karuppasamy, R., & Sivakumar, G. (2020). Experimental investigation of heat transfer enhancement of shell and tube heat exchanger using SnO2-water and Ag-water nanofluids. *Journal of Thermal Science and Engineering Applications, 12*(4), 041016.
42. Goldanlou, A. S., et al. (2021). Heat transfer of hybrid nanofluid in a shell and tube heat exchanger equipped with blade-shape turbulators. *Journal of Thermal Analysis and Calorimetry, 143*(2), 1689–1700.
43. Hu, D., Wang, J., & Tang, Q. (2021). Numerical investigation of a nanofluidic heat exchanger by employing computational fluid dynamic. *Journal of Thermal Analysis and Calorimetry, 144*(5), 1831–1838.
44. Ozden, E., & Tari, I. (2010). Shell side CFD analysis of a small shell-and-tube heat exchanger. *Energy Conversion and Management, 51*(5), 1004–1014.
45. Singh, S. K., & Sarkar, J. (2018). Energy, exergy and economic assessments of shell and tube condenser using hybrid nanofluid as coolant. *International Communications in Heat and Mass Transfer, 98*, 41–48.
46. Shah, R. K., & Sekulic, D. P. (2003). *Fundamentals of heat exchanger design*. Wiley.
47. Guo, J., Cheng, L., & Xu, M. (2009). Optimization design of shell-and-tube heat exchanger by entropy generation minimization and genetic algorithm. *Applied Thermal Engineering, 29*(14–15), 2954–2960.

Chapter 2
Plate Heat Exchanger

Abstract Plate heat exchangers (PHE) are good replacement for shell and tube heat exchanger in food, pharmaceutical and chemical industries due to their high thermal performance. PHE thermal performance can be enhanced by increasing surface area, but this option is limited considering space, pressure losses, pumping power and operational cost. Alternative option is to use fluid with high heat transfer properties. Nanofluids are best heat transfer fluid among the conventional fluids (water and thermal oils). In this chapter, role of nanofluid in enhancing thermal performance of plate heat exchanger has been reviewed. Effect of nanofluid concentration, shape of nanoparticles, Reynold number, geometric parameters of PHE, inlet temperature of nanofluid on effectiveness, heat transfer rate, heat transfer coefficient, pressure drop has been studied.

2.1 Introduction

Energy demand of the world is increasing every single year. To meet growing energy demand, it is crucial to optimize the performance of existing energy systems or implementing new energy systems. In any energy system, heat exchangers (HE) play vital role for achieving the desired energy demand and conversion. Use of heat exchanger type for energy conservation depends on the type of fluid involved in process of heat exchange. For gas to liquid, liquid to liquid heat transfer, shell and tube heat exchangers are used while for gas to gas, gas to liquid, liquid to liquid heat transfer, compact heat exchangers are used. Plate heat exchanger is one of the types of compact heat exchanger that are largely employed in chemical, petroleum, pharmaceutical and food industries due to their ease of sanitation, high thermal efficiency and flexibility [1]. Plate heat exchangers are made of succession of thin flow channels set up between plates. Hot and cold stream flow distribution depends on the perforation of plates and designs of gasket, based on this distribution, plentiful configurations are possible [2].

Improvement in thermal performance of heat exchanger is necessary. Thermal performance of plate heat exchanger can be optimized by enhancing heat transfer area of plate heat exchanger. One possibility is to modify plate surface. Generally, plate

H. M. Ali et al., *Nanofluids for Heat Exchangers*,
https://doi.org/10.1007/978-981-19-3227-4_2

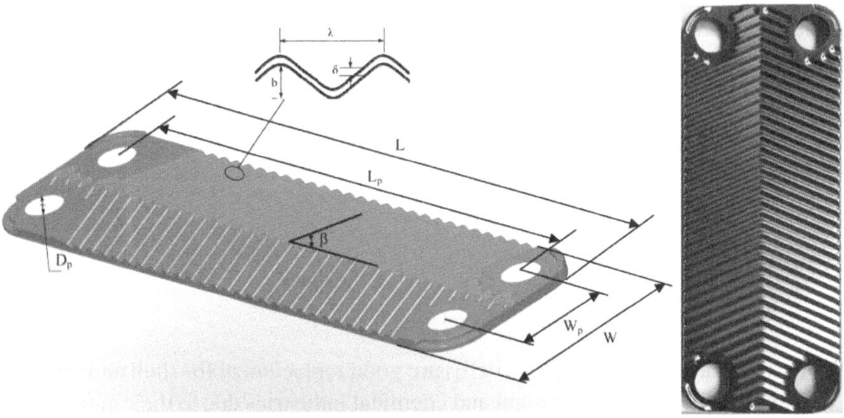

Fig. 2.1 Schematic of chevron corrugated plate [9]

surface is either wavy or corrugated [3]. In chevron corrugated plate heat exchanger, chevron angle β helps in enhancing heat transfer and Nusselt number along with the friction factor while deeper groves helped in development of recirculation zones due to induced core flow velocities [4]. Hessami et al. [5] found that increasing the chevron angle from 45 to 60 helped in changing the flow regime from laminar to turbulent assisting in heat transfer rate. Yang et al. [6] compared geometrical parameters of brazed plate heat exchanger and found that chevron angle is the most effective parameter in enhancing the thermal hydraulic performance of plate heat exchanger. In another study, Nusselt number was enhanced by enhancing chevron angle and Reynold number [7]. Yildiz and Ersoz et al. [8] found good agreement of energy and exergy efficiencies result in chevron plate heat exchanger. Energy and exergy efficiency obtained was 78 and 47.8%, respectively (Fig. 2.1).

Hajabdollahi et al. [10] compared investment and operational cost analysis of gasket plate heat exchanger and shell and tube heat exchanger using genetic algorithm and found 13% decreased investment cost for gasket plate heat exchanger compared to shell and tube heat exchanger but its operational cost was higher than shell and tube heat exchanger. Najafi et al. [11] found that high heat transfer area is required for enhancing the heat transfer rate if pumping power needs to be minimum. It is not possible to achieve high heat transfer at minimum pressure drop.

Besides modifying the design of plate heat exchanger, another way to improve thermal performance of HE is to use fluid with superior thermal properties compared to base fluids. In recent years, nanofluids are considered as next generation heat transfer fluid in heat exchangers due to their superior thermal characteristics compared to base fluids such as water, thermal oils and ethylene glycol.

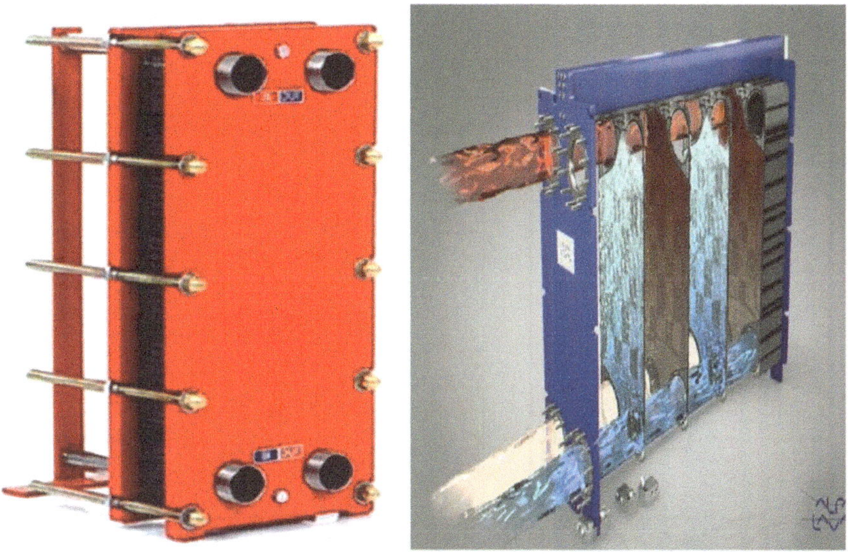

Fig. 2.2 Plate heat exchanger [3]

2.2 Performance of Nanofluid in Plate Heat Exchanger

Effect of nanofluid and different geometries on plate heat exchanger was studied in last decade. Detailed overview of research on plate heat exchangers is given in this study. Shape and gasket of plate heat exchanger is shown in Fig. 2.2.

Pantzali et al. [12] experimentally and numerically studied performance of CuO/H_2O nanofluid on miniature PHE. It was found that maximum heat transfer achieved was up to 115 W using nanofluid at 3 ml/s flow rate and no more than 105 W using water at 11 ml/s. Mandatory volumetric flow rate of nanofluid was 3 times less than that of water. Pantzali et al. [13] inspected the efficacy of nanofluid in PHE as coolants. Result showed that on industrial scale where large amount of nanofluid was required, replacing conventional fluid with nanofluid was seemed unpromising. Pandey and Nema et al. [14] studied ramification of Al_2O_3/H_2O on corrugated plate heat exchanger. Heat transfer characteristics were directly proportional to Reynold and Peclet number and inversely proportional to volume concentration. Also, pumping power was enhanced by enhancing volume concentration because viscosity of nanofluid was increased at higher concentration. 880 W and 111.5% heat transfer and heat transfer coefficient was obtained at 2% volume concentration. Tiwari et al. [15] compared performance of different nanofluids (CeO_2/H_2O, Al_2O_3/H_2O, $TiO_2//H_2O$, SiO_2/H_2O) on counter flow plate heat exchanger. OHTC was enhanced by increasing flow rate of nanofluid because nanoparticle collision with wall surface was enhanced which raised temperature. 35.9, 26.3, 24.1 and 13.9% increment in heat transfer was found at 0.75, 1, 0.75 and 1.25% volume concentration for CeO_2, Al_2O_3, TiO_2 and SiO_2 nanofluid at 3 LPM flow rate, respectively. Geometry of chevron plate is shown in Fig. 2.3.

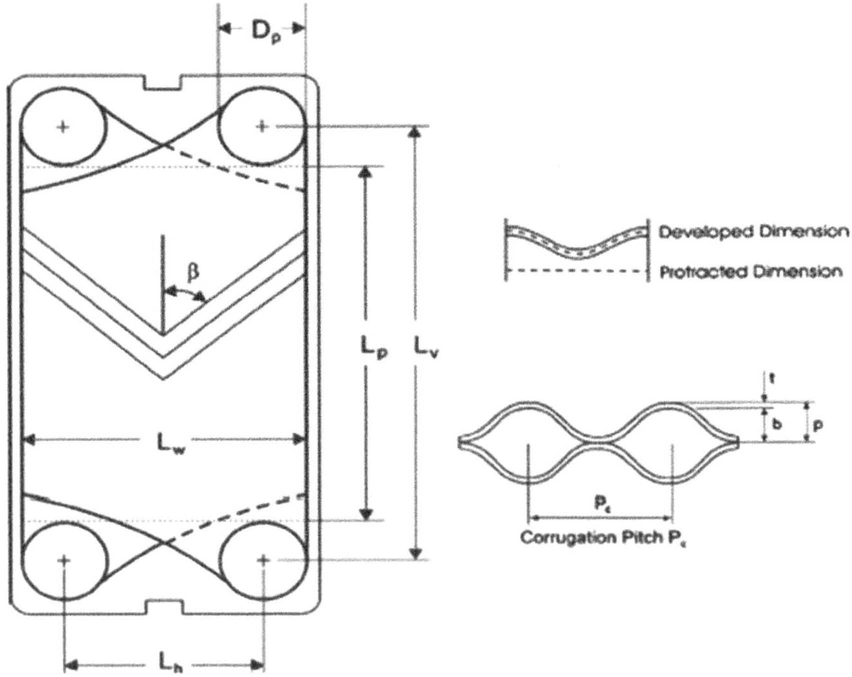

Fig. 2.3 Chevron plate geometric parameters [15]

Tiwari et al. [16] investigated effect of CeO_2/H_2O nanofluid on convective heat transfer coefficient in chevron-type corrugated plate heat exchanger. Convective heat transfer coefficient was enhanced by increasing volume flow rate of coolant and heating fluid, volume concentration and decreasing nanofluid temperature and thermal boundary layer. Pressure drop of nanofluid was equal to that of water. 27.9 and 35.9% enhancement in overall heat transfer coefficient and convective heat transfer coefficient was obtained at 3 LPM flow rate and 0.75% volume concentration. Mare et al. [17] compared performance of CNT and γ-Al_2O_3 nanofluid in plate heat exchanger. Result showed enhancement in convective heat transfer coefficient of about 50 and 42% for CNT and γ-Al_2O_3 in laminar mode. But pressure drop for CNT and γ-Al_2O_3 was 7 times and 3 times more than water. It was suggested that before using nanofluid in heat exchange pressure drop and viscosity of nanofluid must be checked. Goodarzi et al. [18] used MWCNT by functionalization with Cysteine and Ag to raise thermal conductivity in counter flow corrugated plate heat exchanger. Nusselt number and heat transfer coefficient was increased by enhancing weight concentration. Pumping power was also increased at higher concentration because at higher concentration viscosity of fluid and pressure drop were raised. Friction factor was reduced by increasing Peclet number. Khairul et al. [19] enhanced exergy and heat transfer using CuO/H_2O nanofluid in corrugated plate heat exchanger. Friction factor was reduced by increasing Peclet number. Higher pumper power was observed

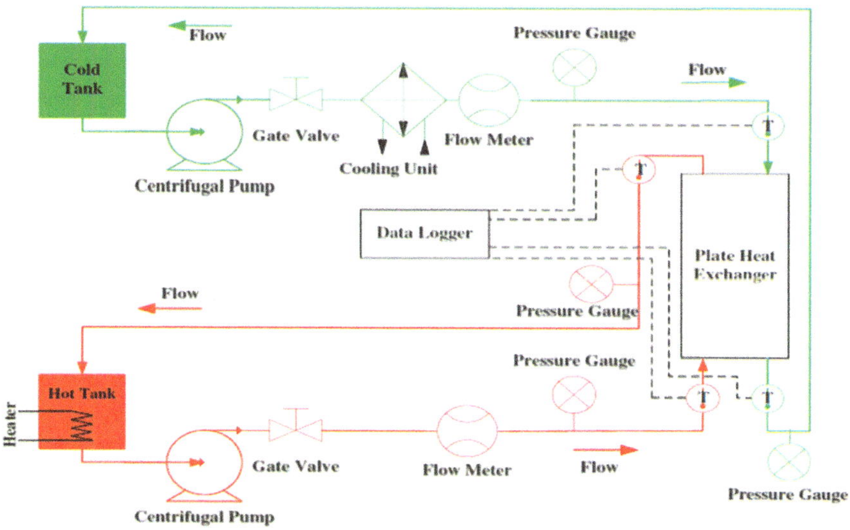

Fig. 2.4 Schematic of plate heat exchanger [19]

at higher volume concentration. About 24.7% increase in heat transfer coefficient was observed at 1.5 vol.%. While exergetic heat transfer effectiveness achieved was 34% at 1.5 vol.%. Schematic of experimental rig used in [19] is given in Fig. 2.4.

Barzegarian et al. [20] suggested TiO_2/H_2O nanofluid as a replacement of base fluid for hot water system. Convective heat transfer coefficient, pressure drop and overall heat transfer coefficient were directly related to Re number and concentration of nanofluid. 8.5, 23.7 and 20% increment in overall heat transfer, convective heat transfer and pressure drop was noticed at 1.5% weight fraction. Tiwari et al. [21] numerically studied corrugated chevron plate heat exchanger using Al_2O_3 and CeO_2/H_2O nanofluid. CeO_2/H_2O performed better as a coolant than Al_2O_3/H_2O and water. Heat transfer rate was improved due to turbulence created by corrugation pattern. Javadi et al. [22] analyzed performance of plate heat exchanger using three different nanofluids (Al_2O_3, SiO_2, TiO_2) with liquid nitrogen as base fluid. Al_2O_3 and TiO_2 showed better thermophysical properties than SiO_2 because of high density of TiO_2 and greater specific heat of Al_2O_3. Pressure drop and entropy generation was enhanced by enhancing volume concentration. SiO_2 showed least entropy generation and pressure drop than other two nanofluids. Prandtl number was increased by decreasing volume concentration (Fig. 2.5).

Huang et al. [23] experimentally investigated effect of MWCNT/DI H_2O and Al_2O_3/DI H_2O on performance of corrugated plate heat exchanger. Corrugated plate is shown in Fig. 2.6. Results showed that heat transfer deteriorated more intensively by MWCNT than Al_2O_3 because of greater MWCNT/H_2O viscosity than Al_2O_3/H_2O. Heat transfer coefficient was reduced by enhancing volume concentration because effect of viscosity was more prominent than effect of thermal conductivity.

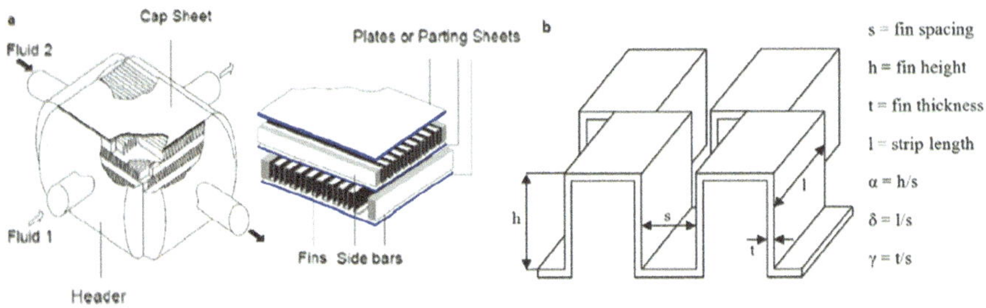

Fig. 2.5 **a** Plate fin heat exchanger schematic. **b** Offset strip fin surface geometry [22]

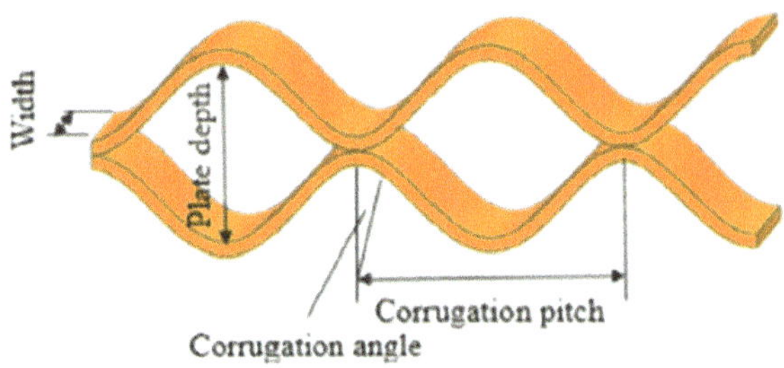

Fig. 2.6 Corrugated plate [23]

Tiwari et al. [24] investigated effect of volume fraction of various nanofluids (CeO_2, Al_2O_3, TiO_2, SiO_2) on plate heat exchanger. Results revealed that above optimum concentration heat transfer rate and overall heat transfer coefficient started to decrease because nanoparticles deposition formed a layer on surface and effective flow area was reduced due to which friction factor and resistance to heat transfer were enhanced. 35.9, 26.3, 24.1 and 13.9% enhancement in heat transfer was noticed at optimum concentration of 0.75, 1, 0.75 and 1.25% for CeO_2/H_2O, Al_2O_3/H_2O, TiO_2/H_2O and SiO_2/H_2O nanofluid, respectively. Huang et al. [25] used hybrid nanofluid (Al_2O_3-MWCNT/H_2O) to study heat transfer properties in chevron plate heat exchanger. Heat transfer coefficient of hybrid nanofluid was greater than Al_2O_3/H_2O. Pressure drop of hybrid nanofluid was less than Al_2O_3/H_2O but slightly greater than H_2O because of enhanced nanoparticles hydrodynamic diameter due to cluster formation in hybrid nanofluid which enhanced viscosity and raised pressure drop. Sarafraz et al. [26] enhanced performance of chevron type flat plate heat exchanger using low frequency vibration and CuO/H_2O nanofluid. Fouling formation on inner wall of heat exchanger was mitigated by using low frequency vibration. Thermal performance of system was improved due to vibration and nanofluid with optimum weight concentration of 0.3%. Raising nanofluid concentration had a greater impact on HTC in turbulent region than in laminar region. Kwon et al. [27] studied different nanofluids (Al_2O_3, ZnO, CNT, TiO_2, Ag) stability and used Al_2O_3 and ZnO due to good stability with low price. Heat transfer coefficient was enhanced up to 30 and 14.5% for Al_2O_3/H_2O and ZnO/H_2O at 6 and 1% volume concentration, respectively. Also, pressure drop was raised up to 84.5 and 11.5% for Al_2O_3 and ZnO at 6 and 1% volume concentration, respectively. Tabari et al. [28] used Peclet number (a better parameter than Re number) instead of Reynold number to analyze performance of nanofluid (TiO_2/H_2O) in industrial plate heat exchanger for milk pasteurization at different concentration. Pressure drop, heat transfer coefficient and Nusselt number showed a direct relation with volume concentration and Peclet number. Heat transfer coefficient was raised to 1251 W/m^{2}K at 0.8% weight concentration and 1000 Pe number. Schematic of experimental rig used for pasteurization of milk is shown in Fig. 2.7.

Sun et al. [29] enhanced stability of nanofluid in plate heat exchanger using SDBS surfactant. Brownian motion of nanofluid was responsible for stable and strong suspension. 669, 645 and 635 W/m^{2}K convective heat transfer coefficient was achieved by utilizing Cu, Al_2O_3 and Fe_2O_3 at 0.3% mass fraction and 1598 Re number. Maximum pressure drop was occurred by Fe_2O_3/H_2O nanofluid.

Kumar et al. [30] analyzed performance of plate heat exchanger at different chevron angle (β = 30°/30°, 30°/60°, 60°/60°) using ZnO/H_2O nanofluid. Heat exchanger with β = 60°/60° performed best in terms of exergy efficiency and heat transfer rate as compared to other chevron angles because of generation of swirl flow

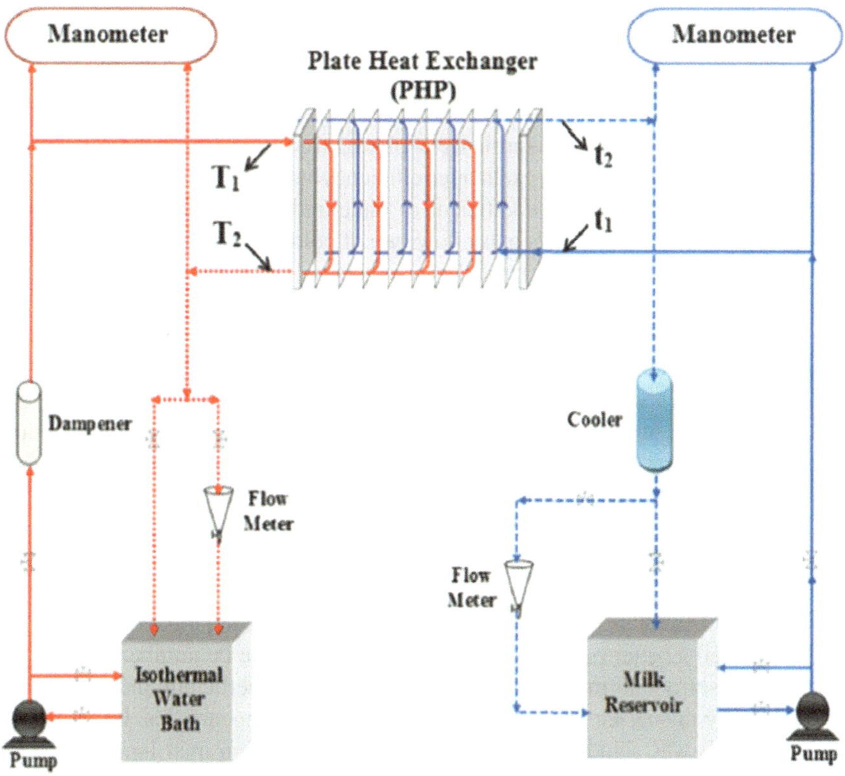

Fig. 2.7 Plate heat exchanger schematic for milk pasteurization [28]

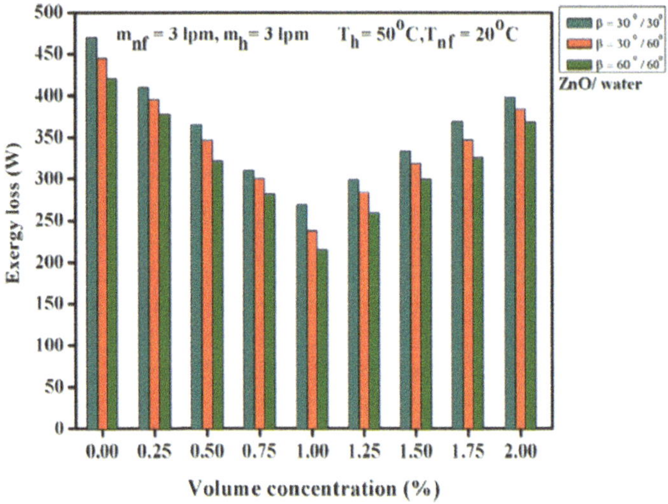

Fig. 2.8 Exergy loss variation for different chevron angles [30]

and increased effective area inside PHE due to increased chevron angle. 26, 32.2 and 27% augmentation in HTRR, HTCR and OHTCR at $\beta = 60°/60°$ and 1% volume concentration, respectively. Figure 2.8 showed exergy loss at different concentration and chevron angles.

Fard et al. [31] compared performance of plate and concentric tube heat exchanger. Results showed that plate heat exchanger efficiency was better than concentric tube heat exchanger. Heat transfer coefficient of plate and concentric tube was 20 and 14% greater than H_2O at 10 g/s mass flow rate of cold and hot fluid. Aliabadi et al. [32] investigated effect of geometrical parameters on performance of plate fin heat exchanger. Results revealed that heat transfer and pressure drop had direct relation with channel length, wing width and wings attack angle, while reverse relation with longitudinal wings pitch, wing height and transfer wings pitch was observed. Figure 2.9 represents vortex generator used in [32].

Ray et al. [33] revealed that mixing HFE-7000 with 1% Al_2O_3 could reduce surface area of heat exchanger by 3.11% and pumping power by 3% on hot side of PHE. Convective heat transfer was enhanced by 4.75% at 0.2 kg s^{-1} flow rate for Al_2O_3 while 4.78% decrement in volumetric flow rate was observed for CuO as compared to EG-H_2O base fluid. Curve fit correlation was proposed for friction factor and Nusselt number given in Table A.2. Tabari and Heris et al. [34] used MWCNT/H_2O nanofluid in Milk Pasteurization Plate heat exchanger. Peclet number was considered a better parameter than Re number because Pe number judged both thermal diffusivity and momentum rates. Heat transfer coefficient and Nu number were raised by raising Pe number and concentration of nanoparticles. Better performance of PHE was achieved under turbulent flow condition than laminar. Kabeel et al. [35] showed that power transmission and effectiveness of PHE was enhanced and reduced by using nanofluid, respectively. High volume flow rate was required in

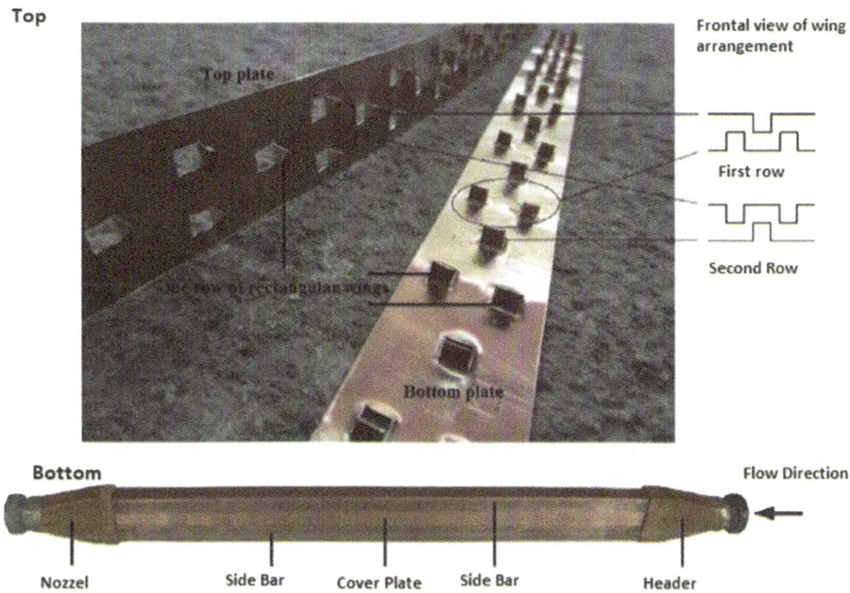

Fig. 2.9 Vortex generator channel [32]

order to compensate high viscosity of nanofluid for achieving constant Re number. Pumping power and HTC was enhanced up to 90 and 13% at 4% nanofluid concentration in comparison with water, respectively. It was observed that use of nanofluid in corrugated plate was doubtful.

Aliabadi et al. [36] analyzed effect of different channels (plain, offset strip, wavy, pin, perforated, louvered and vortex generator) on thermal hydraulic performance in plate fin heat exchanger using Cu/H_2O nanofluid. Thermal hydraulic performance was enhanced by increasing volume flow rate and reducing weight fraction. Maximum pressure drop and maximum surface area was achieved by pin channel and vortex generator, respectively. Aliabadi et al. [37] performed experiment on corrugated wavy (CWC) and parallel plate channel (PPC) plate fin heat exchanger using different nanofluids. CWC enhanced HTC and pressure drop to 44.7 and 39.6% in comparison with PPC. It was observed that Nu number was enhanced by decreasing concentration while friction factor was decreased by decreasing concentration. Area goodness factor was enhanced to 20 and 19.5% using Cu and SiO_2 water based nanofluid. SiO_2/EG-H_2O performed better by reducing mass of EG in base fluid (EG-H_2O). SiO_2/H_2O nanofluid performed best as compared to other nanofluids (Fig. 2.10).

Stogiannis et al. [38] increased heat transfer rate up to 35% using SiO_2/H_2O nanofluid in plate heat exchanger, in comparison with water. Flow rate of nanofluid was 20% lower than water to reduce pressure drop and smaller size equipment. Bhattad et al. [39] showed that discrete phase model was better than homogenous model. Hybrid nanofluid (Al_2O_3-MWCNT/H_2O) enhanced heat transfer coefficient by 39.16% at expense of pumping power. Overall performance index was reduced because increment in pumping power was more than heat transfer rate. Elias et al.

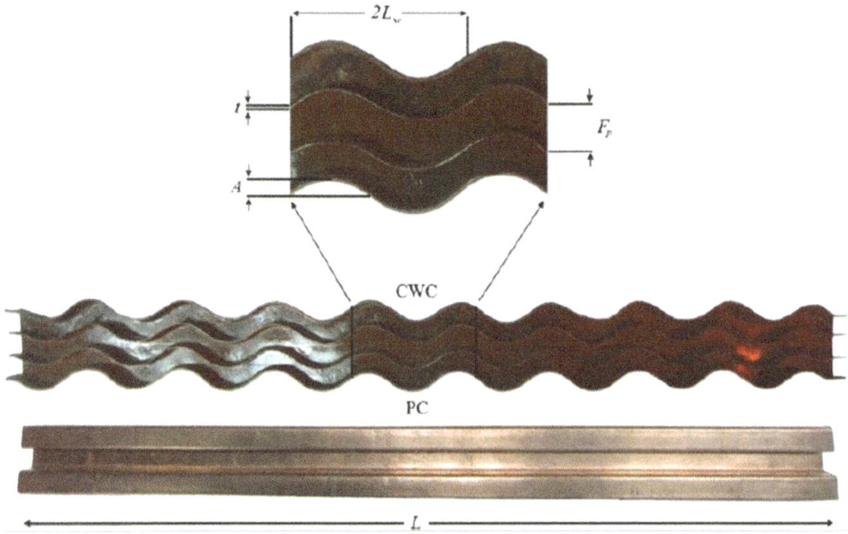

Fig. 2.10 Terminologies and geometry of CWC and PPC [37]

[40] compared performance of Al_2O_3 and SiO_2 water based nanofluid in plate heat exchanger with different chevron angles angle (β=30°/30°, 30°/60°, 60°/60°). Results revealed that Al_2O_3/H_2O performance was superior to SiO_2/H_2O in terms of heat transfer enhancement. Also, heat transfer coefficient was increased by enhancing chevron angle. Wang et al. [41] used polyurethane vessel as an insulator in miniature plate heat exchanger to reduce heat loss. It was suggested that heat exchanger geometry should be redesigned, keeping in mind peculiar properties of nanofluid. Heat transfer was increased to 4% when concentration of graphene nanoplatelets was raised from 0.01 to 0.1%, in comparison with base fluid. Enhancement in viscosity caused ineffectiveness of higher weight concentration of graphene nanofluid in miniature plate heat exchanger.

Sarafraz and Hormozi et al. [42] allowed heat exchanger to run for 720 h continuously to study fouling effect using $MWCNT/H_2O$. It was found that fouling resistance for metal oxides nanofluid was 10 times greater than $MWCNT/H_2O$. At higher operating time (600–720 h), fouling resistance increased to 0.1 and 0.3 at higher concentration of nanofluid (1.5 vol.%). Nusselt number was enhanced by increasing volume concentration (up to optimum value) and inlet temperature of nanofluid. Jokar and Halloran et al. [43] did simulation using Al_2O_3/H_2O in corrugate plate heat exchanger. It was concluded that heat transfer coefficient was reduced by enhancing volume concentration because by increasing concentration, effect of viscosity was more dominant than thermal conductivity. Behrangzade et al. [44] discussed that heat transfer coefficient was greater at low inlet temperature than at higher inlet temperature. Moreover, decrement in friction factor was noticed by enhancing Re number. It was suggested that Ag/H_2O nanofluid should be used at low concentration to improve heat transfer rate with zero increase in pumping power. Safi et al. [45] used

Fig. 2.11 Fabricated and tested WPFs [47]

hybrid nanofluid (MWCNT-TiO_2/H_2O) to study performance of plate heat exchanger. Results showed that heat transfer coefficient ratio was enhanced by increasing volumetric flow rate of nanofluid because effect of nanoparticle collision on inner wall surface was enhanced which resulted in temperature rise. 20.2% increment in HTC was found at 0.08 wt.% and 3.5 LPM flow rate. Abed et al. [46] numerically studied effect of different nanofluids on corrugated trapezoidal PHE. Nu number and pressure drop was enhanced by enhancing trapezoidal channel height, volume concentration and decreasing nanoparticle size and trapezoidal pitch. Performance factor suggested 6 mm longitudinal pitch, 2.5 mm trapezoidal height and CuO/H_2O nanofluid for energy and cost saving. Aliabadi et al. [47] used different passive techniques (perforations, winglets) to enhance performance of wavy plate fin heat exchanger using α-Al_2O_3/H_2O nanofluid. Wavy aspect ratio used was 0.42, 0.33 and 5 mm perforation diameter. Result showed that waviness WPF had greater Nusselt number ratio than perforated WPF (Fig. 2.11).

Zheng et al. [48] compared thermal performance of four different nanofluids (Fe_3O_4/H_2O, Al_2O_3/H_2O, SiC/H_2O, CuO/H_2O) in plate heat exchanger. Fe_3O_4/H_2O showed best thermal performance in terms of overall heat transfer coefficient compared to other nanofluids. Heat transfer enhancement was reduced at higher volume flow rate. For preparation of the four different nanofluids, following dispersant were used for longer stability (Table 2.1).

Table 2.1 Dispersant/surfactant for different nanofluids

Nanofluids	Dispersant	$Mass_P$:$Mass_d$
Fe_3O_4/H_2O	Sodium Citrate	5:1
CuO/H_2O	Sodium hexametaphosphate	4:1
SiC/H_2O	CTAB	5:3
Al_2O_3/H_2O	Sodium hexametaphosphate	5:1

Fazeli et al. [49] developed correlation for convective heat transfer coefficient of water and hybrid nanofluid:

$$h_w = -1492.51822 + 180.86670T_h - 209.53541\dot{V} + 0.247292T_h\dot{V} - 1.74792T_h^2 + 7.50541V^2 \quad (2.1)$$

$$h_{\text{HyNF}} = 11004.17387 - 236.02156T_h - 209.53541\dot{V} - 0.247292T_h\dot{V} + 1.74792T_h^2 + 7.50541V^2 \quad (2.2)$$

Exergy efficiency of PHE using MWCNT-Fe_3O_4/H_2O nanofluid was raised by raising volume concentration of nanofluid and increment in Reynold number as shown in Fig. 2.12. Negative aspect of raising nanofluid concentration is raised in pumping power of the system due to increased viscosity of the fluid at higher concentration. Friction factor was also increased by using nanofluid compared to water. Empirical correlation was developed for friction factor:

$$f = \frac{69.95}{\text{Re}}(1+\varphi)^{-0.42} \quad (2.3)$$

The above equation is valid only for:

$$300 < \text{Re} < 1000 \text{ and } 0 < \varphi < 0.3\%$$

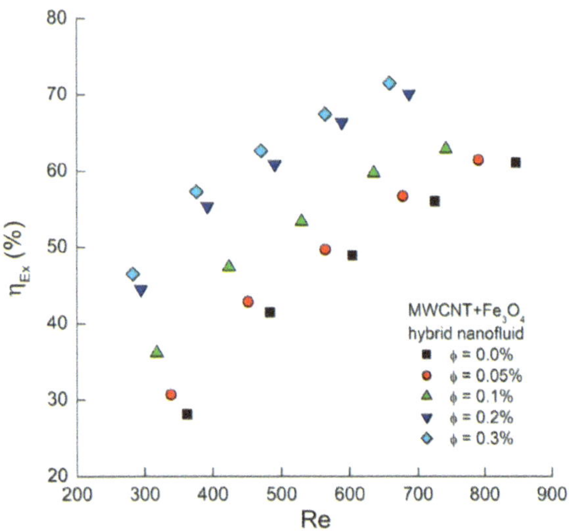

Fig. 2.12 Effect of Re number and volume concentration on the exergy efficiency [50]

[50]	$Nu = 0.1735Re^{0.4655}Pr^{0.4}(1+\varphi)^{0.692}$	$300 < Re < 1000$; $5.5 < Pr < 6.5$; $0 < \varphi < 0.3\%$

Bahiraei et al. [51] numerically compared performance of different shaped alumina nanoparticles as shown in Fig. 2.13 in micro plate heat exchanger.

Numerical results showed that platelets due to higher velocity showed maximum overall heat transfer coefficient, heat transfer rate and pressure loss while Os shaped nanoparticle showed maximum performance index with maximum effectiveness due to minimum pressure drop because of low velocity. It was found that heat transfer rate decreases with increase in inlet temperature of cold fluid (Fig. 2.14; Table 2.2).

Bhattad et al. [52] experimentally compared alumina hybrid nanofluids (SiC, AlN, MWCNT, MgO, CuO) in 4:1 volume ratio. Among the studied nanofluids, Al_2O_3-MWCNT had superior thermal performance compared to other nanofluids with maximum pumping power due to high thermal conductivity and high viscosity. MWCNT has one dimension in micrometer which makes it greater in size causing

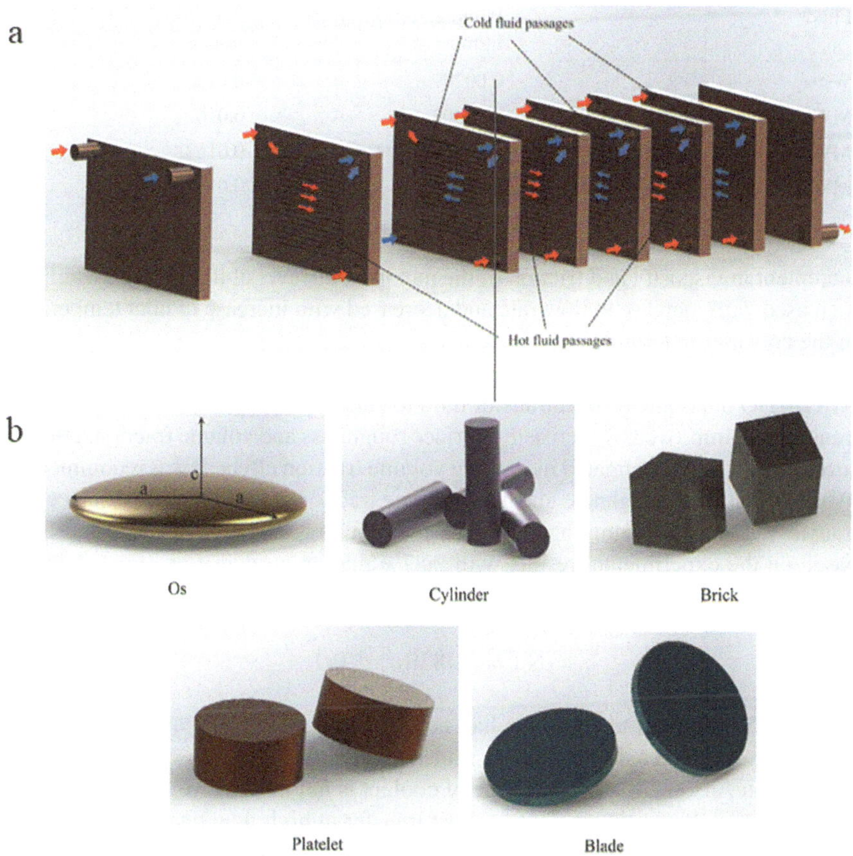

Fig. 2.13 **a** Micro plate heat exchanger schematic. **b** Shapes of alumina nanoparticles [51]

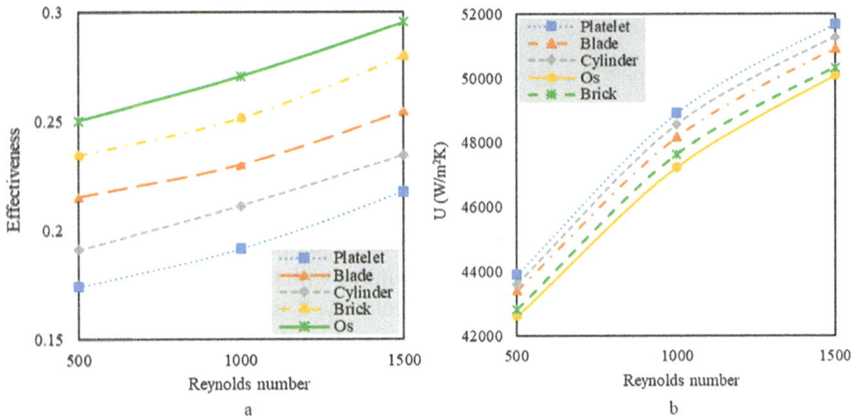

Fig. 2.14 Effect of particle shape on **a** effectiveness and **b** overall heat transfer coefficient [51]

Table 2.2 Pumping power at different volume concentration of nanofluid [50]

Fluid	Pumping power (W) at 3 lit/min	Pumping power (W) at 7 lit/min
Water	0.00289	0.0153
MWCNT-Fe_3O_4/H_2O-0.05 vol.%	0.00303	0.0162
MWCNT-Fe_3O_4/H_2O-0.05 vol.%	0.00314	0.0167
MWCNT-Fe_3O_4/H_2O-0.05 vol.%	0.0032	0.0168

increment in viscosity and friction of the nanofluid. Irreversibility of the system was increased with increase in flow rate and decreased with increase in inlet temperature of the coolant (nanofluid).

Attalla and Maghrabie et al. [53] applied roughness on plate heat exchanger and Al_2O_3/H_2O nanofluid as heat transfer fluid to enhance heat transfer properties. Heat transfer was improved by increasing surface roughness and volume fraction, effect of surface roughness dominated increasing volume fraction effect. It was recommended to use high surface roughness and high-volume fraction at Re = 1000 to enhance heat transfer rate with low pumping power. Correlation for Nusselt number was developed based on the experimental results with 2.87% maximum deviation (Figs. 2.15 and 2.16).

$$\mathrm{Nu} = 0.383\mathrm{Re}^{0.734}\mathrm{Pr}^{\frac{1}{3}} \tag{2.4}$$

Increasing coolant temperature had negative impact on thermal performance of plate heat exchanger due to decrease in temperature difference causing low heat transfer rate, also increasing flow rate of coolant is not ideal as well because of very less time available to the coolant for heat transfer at high flow rates. No significant improvement in heat transfer was observed when Al_2O_3-graphene nanofluid was used

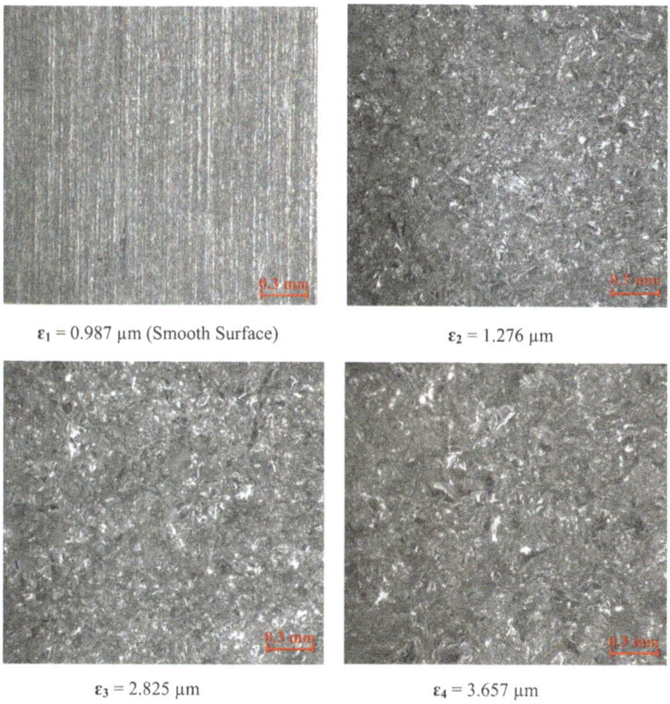

Fig. 2.15 Surface roughness used in plate heat exchanger [53]

in 4:1 amount compared to Al_2O_3/H_2O because of low concentration of graphene nanoparticles used [54].

2.3 Thermal Analysis of Plate Heat Exchanger

Effectiveness of counter flow heat exchanger is evaluated using following relation [55]:

$$\varepsilon = \frac{1 - e^{-NTU(1-C^*)}}{1 - C^* e^{-NTU(1-C^*)}} \quad (2.5)$$

where C^* is the heat capacity rate ratio and is given as:

$$C^* = \frac{C_{min}}{C_{max}} \quad (2.6)$$

Overall heat transfer coefficient of heat exchanger can be evaluated using following:

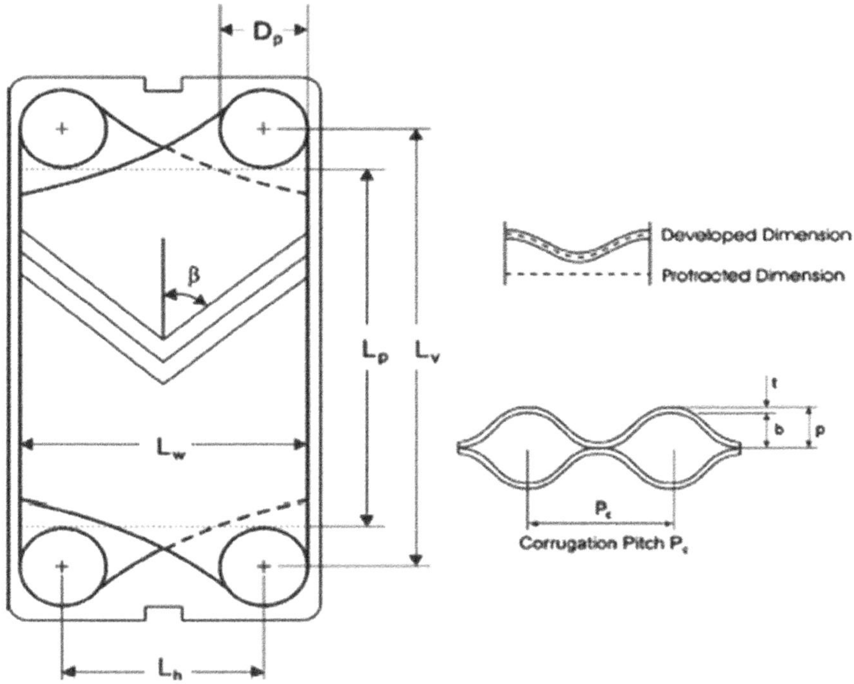

Fig. 2.16 Chevron plate geometric parameters [15]

$$\frac{1}{U} = R_{\mathrm{f,\,h}} + R_{\mathrm{f,\,c}} + \left[\frac{1}{h_{\mathrm{h}}} + \frac{1}{h_{\mathrm{c}}}\right] + \left(\frac{t}{k}\right)_{\mathrm{w}} \tag{2.7}$$

where w, t and k represents the wall of heat exchanger, thickness and conductivity, respectively.

Diameter of chevron plate can be determined using:

$$d_e = 2b \tag{2.8}$$

b is the mean flow channel gap:

$$b = p - t \tag{2.9}$$

$$p = \frac{L_c}{N_t} \tag{2.10}$$

Hydraulic diameter can be expressed as [11]:

$$D_{\mathrm{e}} = \frac{4(\text{channel flow area})}{\text{wetted surface}} = \frac{4A_{\mathrm{c}}}{P_{\mathrm{w}}} = \frac{4bL_{\mathrm{w}}}{2(b + L_{\mathrm{w}}\phi)} = \frac{2b}{\phi} \tag{2.11}$$

Reynold number in plate heat exchanger is given as:

$$\mathrm{Re} = \frac{G_{\mathrm{ch}} \times d_{\mathrm{e}}}{\mu} \tag{2.12}$$

where G_{ch} is mass flux for each channel:

$$G_{\mathrm{ch}} = \frac{\dot{m}_{\mathrm{ch}}}{A_{\mathrm{ch}}} \tag{2.13}$$

$$A_{\mathrm{ch}} = b \times L_{\mathrm{w}} \tag{2.14}$$

$$L_{\mathrm{w}} = L_{\mathrm{h}} + d_{\mathrm{p}} \tag{2.15}$$

$$L_{\mathrm{w}} = L_{\mathrm{h}} + d_{\mathrm{p}} \quad (2.15)$$

$\dot{m}_{\mathrm{ch}}$ is the mass flow rate in plate heat exchanger channel:

$$\dot{m}_{\mathrm{ch}} = \frac{\dot{m}}{n_{\mathrm{cp}}} \tag{2.16}$$

$$n_{\mathrm{cp}} = \frac{n_{\mathrm{t}} - 1}{2n_{\mathrm{p}}} \tag{2.17}$$

Pressure Drop in Plate Heat Exchanger

Pressure drop in PHE is sum of pressure drop due to port and pressure drop due to channel.

Channel Pressure Drop

Pressure drop through channels can be evaluated using following:

$$\Delta P = 4f \frac{L_{\mathrm{eff}} N_{\mathrm{p}}}{D_{\mathrm{e}}} \frac{G_{\mathrm{c}}^2}{2\rho} \left(\frac{\mu_{\mathrm{b}}}{\mu_{\mathrm{w}}} \right)^{-0.17} \tag{2.18}$$

where f is the friction coefficient.

Port Pressure Drop

$$\Delta P_{\mathrm{p}} = \frac{1.4 N_{\mathrm{p}} G_{\mathrm{p}}^2}{2\rho} \tag{2.19}$$

$$G_{\mathrm{p}} = \frac{\dot{m}}{\frac{\pi D_{\mathrm{p}}^2}{4}} \tag{2.20}$$

Entropy Generation in Plate Heat Exchanger

Entropy generation in PHE is due to thermal and friction factor entropy generation:

$$S_{\text{gen}} = \dot{S}_{\text{gen,T}} + \dot{S}_{\text{gen,F}} \tag{2.21}$$

Thermal entropy generation can be evaluated as:

$$\dot{S}_{\text{gen,T}} = \frac{\dot{Q}_{\text{avg}}}{\text{Nu Re Pr } T_{\text{c,i}} T_{\text{c,o}} L} \tag{2.22}$$

$$\dot{S}_{\text{gen,F}} = \frac{8 f \dot{m}^3 L}{\rho^2 \pi^2 D_h^5 \left(T_{\text{c,o}} - T_{\text{c,i}}\right)} \ln\left(\frac{T_{\text{c,o}}}{T_{\text{c,i}}}\right) \tag{2.23}$$

2.4 Discussion

Figure 2.17 shows maximum heat transfer rate as a function of flow rate attained in different literature. Heat transfer rate increases by increasing flow rate because at higher flow rates, turbulent effect increases which allows the better mixing of particles in fluid resulting in increase in heat transfer rate. Fazeli et al. and Alklaibi et al. [49, 50] analyzed heat transfer rate using CuO/H_2O nanofluid with 4% volume

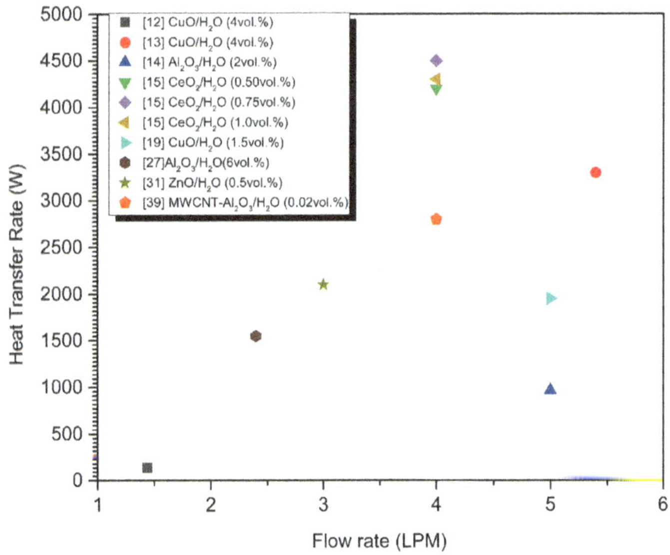

Fig. 2.17 Result comparison of heat transfer rate versus flow rate in literature

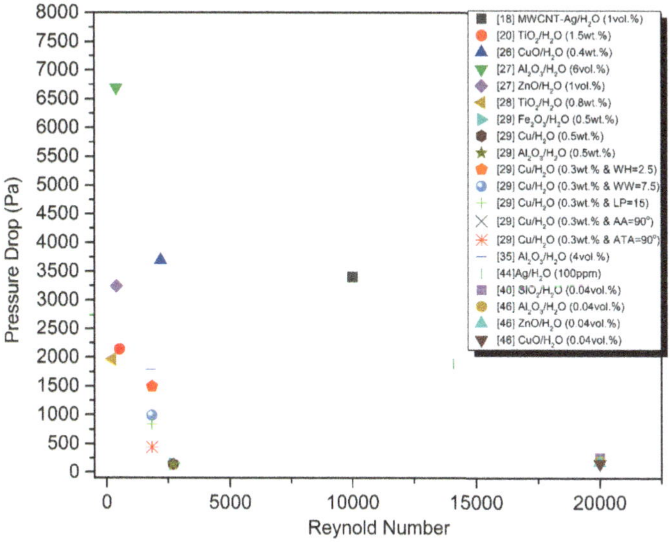

Fig. 2.18 Variation of pressure drop against different Reynold Number using different nanofluids

concentration at two different flow rates. Alklaibi et al. [50] achieved heat transfer rate of 3500 W at 5.5 LPM while heat transfer achieved by [49] was 145 W at 0.96 LPM. Maximum heat transfer rate was achieved at 4 LPM using CeO_2/H_2O nanofluid with 0.75% concentration as compared to other nanofluids and flow rate given in Fig. 2.17. Moreover, it can also be seen that binary nanofluid MWCNT-Al_2O_3/H_2O [39] had almost 37% less heat transfer than CeO_2/H_2O [53] at 4 LPM flow rate due to stability problem of binary nanofluid.

Figure 2.18 showed variation in pressure drop at different Reynold number for different nanofluids. From Fig. 2.18, it was observed that maximum pressure drop occurred was in laminar flow at 6% volume concentration of Al_2O_3/H_2O [27] as compared to other nanofluids. This was due to higher concentration of nanofluid which caused enhancement in viscosity of nanofluid thus resulting enhancement in friction factor and pressure drop. Minimum pressure drop was achieved for [29] and [46] in turbulent flow condition with low concentration of nanofluids used (0.5 wt.% and 0.04 vol.%). Low pressure drop for [29] was due to low viscosity and friction factor of nanofluid which was mainly due to low concentration whereas [46] was a numerical study therefore, experimental results were required to validate numerical results. It was concluded that most of research was conducted with Reynold number ranging from 0 to 5000 (Table 2.3).

Table 2.3 Summary of literature on the usage of nanofluid in plate heat exchanger

References	Heat exchanger	Nanofluid	Nanoparticle size	Surfactant	Concentration	Flow	Outcomes
[12]	Miniature plate	CuO/H_2O	30 nm	–	4 vol.%		Maximum heat transfer achieved was up to 115 W using nanofluid at 3 ml/s flow rate
[13]	Commercial herringbone-type plate	CuO/H_2O	30 nm	CTAB	4 vol.%	Turbulent	Replacing base fluid with nanofluid was seemed unpromising on industrial scale where large amount of nanofluid was required
[14]	Counter flow corrugated plate	Al_2O_3/H_2O	40–50 nm	–	2–4 vol.%	Turbulent	880 W and 111.5% heat transfer and heat transfer coefficient was obtained at 2% volume concentration
[15]	Counter flow plate	CeO_2, Al_2O_3, TiO_2, SiO_2 /H_2O	30, 45, 10, 10 nm	CTAB	0.5–3 vol.%		CeO_2/H_2O nanofluid performed best with 35.9% heat transfer improvement in comparison with other nanofluids

(continued)

Table 2.3 (continued)

References	Heat exchanger	Nanofluid	Nanoparticle size	Surfactant	Concentration	Flow	Outcomes
[16]	Chevron-type corrugated plate	CeO_2/H_2O	30 nm		0.5–3 vol.%	Turbulent	35.9% increment in convective heat transfer coefficient was achieved at 0.75% volume concentration
[17]	Plate heat exchangers	γ-Al_2O_3, CNT/H_2O	37, 9–10 nm	Used but not given	1 and 0.55 vol.%	Laminar	50 and 42% enhancement in convective heat transfer coefficient for CNT and γ-Al_2O_3 nanofluid
[18]	Counter flow corrugated plate	MWCNT-Ag/H_2O	10–20 nm	GA	0–1 wt.%	Turbulent	41.3073% increment in heat transfer coefficient was found at 1 wt.% of hybrid nanofluid
[19]	Corrugate plate	CuO/H_2O	–	–	0.5–1.5 vol.%	Turbulent	34% exergy was achieved at 1.5% volume concentration
[20]	Brazed plate heat exchanger	TiO_2/H_2O	20 nm	CTAB	0.3–1.5 wt.%	Turbulent	8.5, 23.7 and 20% increment in overall heat transfer, convective heat transfer and pressure drop was noticed at 1.5% weight fraction

(continued)

Table 2.3 (continued)

References	Heat exchanger	Nanofluid	Nanoparticle size	Surfactant	Concentration	Flow	Outcomes
[21]	Counter flow chevron corrugated plate	CeO_2, Al_2O_3/H_2O			0.2–3 vol.%	Turbulent	At same pressure drop, nanofluid provided more heat transfer than base fluid, which reduced pumping cost
[22]	Plate heat exchanger	Al_2O_3, SiO_2, TiO_2/Liquid Nitrogen	–	–	0.2–2 vol.%	–	57 and 50% enhancement in entropy generation ratio was observed for TiO_2 and Al_2O_3, respectively
[23]	Chevron plate	MWCNT, Al_2O_3 /DI H_2O	9.5 nm MWCNT, 40 nm Al_2O_3	SDBS	2.18–10.36 wt.% Al_2O_3, 0.02–0.1 wt.%	Turbulent	20% enhancement in viscosity was noticed for MWCNT/H_2O nanofluid at 0.0555 vol.%
[24]	Commercial plate	CeO_2, Al_2O_3, TiO_2, SiO_2/H_2O	30, 45, 10, 10 nm	CTAB	0–3 vol.%	Turbulent	Highest performance index was improved by CeO_2/H_2O at 0.75 vol.% which was 9.5%
[25]	Chevron plate	Al_2O_3-MWCNT/H_2O	40 nm Al_2O_3	Not used	1:2.5 vol. ratio		Pressure drop of hybrid nanofluid was less than Al_2O_3/H_2O but slightly greater than H_2O

(continued)

Table 2.3 (continued)

References	Heat exchanger	Nanofluid	Nanoparticle size	Surfactant	Concentration	Flow	Outcomes
[26]	Chevron-type flat plate	CuO/H_2O	50 nm	SDS	0.1–0.4 wt.%	Laminar and Turbulent	Fouling of nanoparticles was minimized by enhancing the frequency from 30 to 60 Hz at 0.3 wt.%
[27]	Plate heat exchanger	Al_2O_3, ZnO/H_2O	20–50 nm Al_2O_3, 20–50 nm ZnO		1–6 vol.%	Laminar	HTC was enhanced up to 30 and 14.5% for Al_2O_3/H_2O and ZnO/H_2O at 6 and 1% nanoparticle concentration, respectively
[28]	Plate heat exchanger	TiO_2/H_2O	10–15 nm		0.25–0.8 wt.%	Laminar and Turbulent	Heat transfer coefficient was raised to 1251 W/m^2K at 0.8% weight concentration and 1000 Pe number
[29]	Plate heat exchanger	Cu, Fe_2O_3, Al_2O_3/H_2O	50 nm	SDBS	0.1–0.5 wt.%	Laminar-Turbulent	Al_2O_3/H_2O was more stable than other two nanofluids used

(continued)

Table 2.3 (continued)

References	Heat exchanger	Nanofluid	Nanoparticle size	Surfactant	Concentration	Flow	Outcomes
[30]	Chevron angle plate	ZnO/H_2O		CTAB	0.5–2 vol.%	Turbulent	26, 32.2 and 27% augmentation in HTRR, HTCR and OHTCR at β=60°/60° and 1% volume concentration, respectively
[31]	Plate and concentric tube	ZnO/H_2O			0.5 vol.%		Heat transfer coefficient of plate and concentric tube was 20 and 14 greater than water at 10 g/s mass flow rate of cold and hot fluid
[32]	Plate fin	Cu/H_2O	40 nm		0–0.3 wt.%	Laminar	New correlation was developed for fanning friction factor and Colburn factor of nanofluid for flow inside vortex generator
[33]	Minichannel plate	Al_2O_3, SiO_2, $CuO/EG\text{-}H_2O$ (60:40)	45 nm		0.5 and 1% vol.%		Mixing HFE-7000 with Al_2O_3 could reduce surface area of heat exchanger by 3.11% and pumping power by 3% on hot side of PHE

(continued)

Table 2.3 (continued)

References	Heat exchanger	Nanofluid	Nanoparticle size	Surfactant	Concentration	Flow	Outcomes
[34]	PHE	MWCNT/H_2O	5–10 nm	GA	0.25–0.55 wt.%	Laminar and Turbulent	At 1000 Pe number and 0.55 wt.%, Nu number achieved was 12.66
[35]	Corrugate plate	Al_2O_3/H_2O	47 nm	No surfactant used	1–4 vol.%	Laminar	Pumping power and HTC was enhanced up to 90 and 13% at 4% nanofluid concentration in comparison with water, respectively
[36]	Plate fin	Cu/H_2O	30–50 nm	–	0.1–0.4 wt.%		Pressure drop was enhanced by 9.7% at 4 LPM and 0.4 wt.%, while HTC was enhanced by 21.9% at 3.5 LPM and 0.4 wt.%, for plain channel
[37]	Wavy channel plate fin	Cu, Al_2O_3, SiO_2, ZnO, TiO_2, Fe_2O_3, CuO/DI H_2O	30–40 nm (Cu)	No surfactant	0.1–0.4 wt.% and 0.1 wt.% for oxide nanoparticles		Area goodness factor was enhanced to 20 and 19.5% using Cu and SiO_2 water based nanofluid

(continued)

Table 2.3 (continued)

References	Heat exchanger	Nanofluid	Nanoparticle size	Surfactant	Concentration	Flow	Outcomes
[38]	Plate heat exchanger (undulated surface)	SiO_2/H_2O		Tetrasodium Pyrophosphate	1 and 2 vol.%	Laminar	Heat transfer rate was raised up to 35% using SiO_2/H_2O nanofluid in plate heat exchanger, in comparison with water
[39]	Commercial plate	MWCNT-Al_2O_3/H_2O			0.01–0.03 vol.%	Turbulent	Hybrid nanofluid (Al_2O_3-MWCNT/H_2O) enhanced heat transfer coefficient by 39.16% at expense of pumping power
[40]	Chevron plate	SiO_2, Al_2O_3/H_2O	Not given	Not given	0–1 vol.%		Al_2O_3 heat transfer coefficient was 0.67 and 2.1% greater than SiO_2/H_2O and water, respectively
[41]	Miniature plate	Graphene/EG-H_2O	2 nm	Triton X-100	0.01–1 wt.%	Laminar	Enhancement in viscosity caused ineffectiveness of higher weight concentration of graphene nanofluid in miniature plate heat exchanger

(continued)

Table 2.3 (continued)

References	Heat exchanger	Nanofluid	Nanoparticle size	Surfactant	Concentration	Flow	Outcomes
[42]	Chevron-type plate	$MWCNT/H_2O$		Nonylphenolethoxilate	0.5–1.5 vol.%	Laminar. Transient, Turbulent	Thermal conductivity coefficient was enhanced up to 68% by using $MWCNT/H_2O$ nanofluid
[43]	Corrugated plate	Al_2O_3/H_2O			1–4 vol.%	Laminar Turbulent	Heat transfer coefficient was reduced by enhancing volume concentration
[44]	Gasket plate	Ag/H_2O	–	–	100 ppm	–	Overall heat transfer was improved from 6.18 to 16.79% using Ag/H_2O nanofluid
[45]	Commercial plate	$MWCNT\text{-}TiO_2/H_2O$	2.3 nm	Not given	0.02–0.08 wt.%		20.2% increment in HTC was found at 0.08 wt.% and 3.5 LPM flow rate
[46]	Corrugated trapezoidal plate	Al_2O_3, SiO_2, CuO, ZnO	20–80 nm		0–4 vol.%	Laminar and Turbulent	Performance factor showed that 6 mm longitudinal pitch, 2.5 mm trapezoidal height and CuO/H_2O nanofluid was suggested for energy and cost saving

(continued)

Table 2.3 (continued)

References	Heat exchanger	Nanofluid	Nanoparticle size	Surfactant	Concentration	Flow	Outcomes
[47]	Wavy plate-fin heat exchanger	α-Al_2O_3/H_2O	50 nm	No surfactant used			Waviness WPF had greater Nusselt number ratio than perforated WPF
[48]	Plate heat exchanger	Al_2O_3, SiC, CuO, Fe_3O_4/H_2O			0.05–1 wt.%		Heat transfer coefficient was enhanced by 30.8% using Fe_3O_4/H_2O nanofluid at 1 wt.% concentration and 8 L/min flow rate
[49]	Brazed plate heat exchanger	MWCNT-CuO/H_2O	5–15 nm and 40 nm	GA and Tween-80	0.05 wt.%		Maximum enhancement of 139% was observed in convective heat transfer coefficient at 24.4 L/min flow rate of nanofluid with 35 °C and 6 L/min flow rate of hot fluid
[50]	Corrugated plate heat exchanger	MWCNT-Fe_3O_4	10–20 nm		0.05–0.3 vol.%	Laminar	NTU and effectiveness for nanofluid was 24.3 and 4.53% greater than DI water at 0.3 and 0.05 vol.%, respectively

(continued)

Table 2.3 (continued)

References	Heat exchanger	Nanofluid	Nanoparticle size	Surfactant	Concentration	Flow	Outcomes
[51]	Micro plate heat exchanger	Boehmite Alumina/H_2O-EG			1 vol.%	Laminar	Os shaped had maximum performance index followed by bricks, cylinder, blade and platelet shaped nanoparticles
[52]	Corrugated plate heat exchanger	Al_2O_3-SiC, AlN, MgO, CuO, MWCNT/H_2O		CTAB and SDS	s		Heat transfer ratio, performance index and pumping power was enhanced by 31.2, 12.46 and 0.08% using Al_2O_3-MWCNT/H_2O nanofluid, respectively
[53]	Roughened plate heat exchanger	Al_2O_3/H_2o			1.2–2.6 vol.%	Laminar and Turbulent	Nu number was improved by 30.8, 18, 12, and 8% by using 3.657, 2.825, 1.276 and 0.378 μm surface roughness at 2.6 volume fraction of nanofluid
[54]	Corrugated plate heat exchanger	Al_2O_3-Graphene/H_2O		CTAB	0.01 vol.%		Heat transfer rate and heat transfer coefficient was enhanced by 5.34 and 25.36% using hybrid nanofluid

References

1. Shah, R. K., & Sekulic, D. P. (2003). *Fundamentals of heat exchanger design*. Wiley.
2. Gut, J. A., & Pinto, J. M. (2004). Optimal configuration design for plate heat exchangers. *International Journal of Heat and Mass Transfer, 47*(22), 4833–4848.
3. Yazıcıoğlu, A. G., et al. (2018). 4.2 heat exchangers. In I. Dincer (Ed.), *Comprehensive energy systems* (pp. 40–69). Elsevier, Oxford.
4. Muley, A., & Manglik, R. M. (1997). Enhanced heat transfer characteristics of single-phase flows in a plate heat exchanger with mixed chevron plates. *Journal of Enhanced Heat Transfer, 4*(3).
5. Li, Y., Hessami, M.-A., & Hu, E. J. (1998). Experimental investigation of heat transfer and pressure drop in cross-corrugated plate heat exchangers. In *International Conference on Energy and Environment*. China Machine Press.
6. Yang, J., Jacobi, A., & Liu, W. (2017). Heat transfer correlations for single-phase flow in plate heat exchangers based on experimental data. *Applied Thermal Engineering, 113*, 1547–1557.
7. Khan, T., et al. (2010). Experimental investigation of single phase convective heat transfer coefficient in a corrugated plate heat exchanger for multiple plate configurations. *Applied Thermal Engineering, 30*(8–9), 1058–1065.
8. Yildiz, A., & Ersöz, M. A. (2015). Theoretical and experimental thermodynamic analyses of a chevron type heat exchanger. *Renewable and Sustainable Energy Reviews, 42*, 240–253.
9. Zhang, J., et al. (2019). A review of heat transfer enhancement techniques in plate heat exchangers. *Renewable and Sustainable Energy Reviews, 101*, 305–328.
10. Hajabdollahi, H., Naderi, M., & Adimi, S. (2016). A comparative study on the shell and tube and gasket-plate heat exchangers: The economic viewpoint. *Applied Thermal Engineering, 92*, 271–282.
11. Najafi, H., & Najafi, B. (2010). Multi-objective optimization of a plate and frame heat exchanger via genetic algorithm. *Heat and Mass Transfer, 46*(6), 639–647.
12. Pantzali, M., et al. (2009). Effect of nanofluids on the performance of a miniature plate heat exchanger with modulated surface. *International Journal of Heat and Fluid Flow, 30*(4), 691–699.
13. Pantzali, M., Mouza, A., & Paras, S. J. C. E. S. (2009). Investigating the efficacy of nanofluids as coolants in plate heat exchangers (PHE). *Chemical Engineering Science, 64*(14), 3290–3300.
14. Pandey, S. D., Nema, V. J. E. T., & Science, F. (2012). Experimental analysis of heat transfer and friction factor of nanofluid as a coolant in a corrugated plate heat exchanger. *Experimental Thermal and Fluid Science, 38*, 248–256.
15. Tiwari, A. K., et al. (2013). Performance comparison of the plate heat exchanger using different nanofluids. *Experimental Thermal and Fluid Science, 49*, 141–151.
16. Tiwari, A. K., Ghosh, P., & Sarkar, J. J. A. T. E. (2013). Heat transfer and pressure drop characteristics of CeO_2/water nanofluid in plate heat exchanger. *Applied Thermal Engineering, 57*(1–2), 24–32.
17. Maré, T., et al. (2011). Comparison of the thermal performances of two nanofluids at low temperature in a plate heat exchanger. *Experimental Thermal and Fluid Science, 35*(8), 1535–1543.
18. Goodarzi, M., et al. (2015). Investigation of heat transfer and pressure drop of a counter flow corrugated plate heat exchanger using MWCNT based nanofluids. *International Communications in Heat and Mass Transfer, 66*, 172–179.
19. Khairul, M. A., et al. (2014). Heat transfer performance and exergy analyses of a corrugated plate heat exchanger using metal oxide nanofluids. *International Communications in Heat and Mass Transfer, 50*, 8–14.
20. Barzegarian, R., et al. (2016). Experimental investigation on heat transfer characteristics and pressure drop of BPHE (brazed plate heat exchanger) using TiO_2–water nanofluid. *74*, 11–18.
21. Tiwari, A. K., et al. (2014). Numerical investigation of heat transfer and fluid flow in plate heat exchanger using nanofluids. *International Journal of Thermal Sciences, 85*, 93–103.

22. Javadi, F. S., et al. (2013). The effects of nanofluid on thermophysical properties and heat transfer characteristics of a plate heat exchanger. *International Communications in Heat and Mass Transfer, 44*, 58–63.
23. Huang, D., et al. (2015). Pressure drop and convective heat transfer of Al_2O_3/water and MWCNT/water nanofluids in a chevron plate heat exchanger. *International Journal of Heat and Mass Transfer, 89*, 620–626.
24. Tiwari, A. K., et al. (2015). Particle concentration levels of various nanofluids in plate heat exchanger for best performance. *International Journal of Heat and Mass Transfer, 89*, 1110–1118.
25. Huang, D., et al. (2016). Effects of hybrid nanofluid mixture in plate heat exchangers. *Experimental Thermal and Fluid Science, 72*, 190–196.
26. Sarafraz, M., et al. (2017). Low-frequency vibration for fouling mitigation and intensification of thermal performance of a plate heat exchanger working with CuO/water nanofluid. *Applied Thermal Engineering, 121*, 388–399.
27. Kwon, Y., et al. (2011). Heat transfer and pressure drop characteristics of nanofluids in a plate heat exchanger. *Journal of Nanoscience and Nanotechnology, 11*(7), 5769–5774.
28. Taghizadeh-Tabari, Z., et al. (2016). The study on application of TiO_2/water nanofluid in plate heat exchanger of milk pasteurization industries. *Renewable and Sustainable Energy Reviews, 58*, 1318–1326.
29. Sun, B., et al. (2016). Investigation on the flow and convective heat transfer characteristics of nanofluids in the plate heat exchanger. *Experimental Thermal and Fluid Science, 76*, 75–86.
30. Kumar, V., et al. (2016). Effect of chevron angle on heat transfer performance in plate heat exchanger using ZnO/water nanofluid. *Energy Conversion and Management, 118*, 142–154.
31. Fard, M. H., Talaie, M. R., & Nasr, S. J. T. (2011). Numerical and experimental investigation of heat transfer of ZnO/water nanofluid in the concentric tube and plate heat exchangers. 15(1).
32. Khoshvaght-Aliabadi, M., Hormozi, F., & Zamzamian, A. J. A. T. E. (2014). Effects of geometrical parameters on performance of plate-fin heat exchanger: Vortex generator as core surface and nanofluid as working media. *70*(1), 565–579.
33. Ray, D. R., et al. (2014). Experimental and numerical investigations of nanofluids performance in a compact minichannel plate heat exchanger. *International Journal of Heat and Mass Transfer, 71*, 732–746.
34. Tabari, Z. T., & Heris, S. Z. J. (2015). Heat transfer performance of milk pasteurization plate heat exchangers using MWCNT/water nanofluid. *Journal of Dispersion Science and Technology, 36*(2), 196–204.
35. Kabeel, A., El Maaty, T. A., & El Samadony, Y. J. A. T. E. (2013). The effect of using nanoparticles on corrugated plate heat exchanger performance. 52(1), 221–229.
36. Khoshvaght-Aliabadi, M., et al. (2014). Experimental analysis of thermal–Hydraulic performance of copper–water nanofluid flow in different plate-fin channels. *52*, 248–258.
37. Khoshvaght-Aliabadi, M., et al. (2014). Wavy channel and different nanofluids effects on performance of plate-fin heat exchangers. *28*(3), 474–484.
38. Stogiannis, I., Mouza, A., & Paras, S. J. I. (2015). Efficacy of SiO_2 nanofluids in a miniature plate heat exchanger with undulated surface. *92*, 230–238.
39. Bhattad, A., et al. (2018). Discrete phase numerical model and experimental study of hybrid nanofluid heat transfer and pressure drop in plate heat exchanger. *International Communications in Heat and Mass Transfer, 91*, 262–273.
40. Elias, M., et al. (2014). Performance investigation of a plate heat exchanger using nanofluid with different chevron angle. In *Advanced materials research*. Trans Tech Publications
41. Wang, Z., et al. (2018). Experimental comparative evaluation of a graphene nanofluid coolant in miniature plate heat exchanger. *International Journal of Thermal Sciences, 130*, 148–156.
42. Sarafraz, M., Hormozi, F. J. E. T., & Science, F. (2016). Heat transfer, pressure drop and fouling studies of multi-walled carbon nanotube nano-fluids inside a plate heat exchanger. *72*, 1–11.
43. Jokar, A., et al. (2013). Heat transfer and fluid flow analysis of nanofluids in corrugated plate heat exchangers using computational fluid dynamics simulation. *5*(1), 011002.

44. Behrangzade, A., & Heyhat, M. M. J. A. T. E. (2016). The effect of using nano-silver dispersed water based nanofluid as a passive method for energy efficiency enhancement in a plate heat exchanger. *102*, 311–317.
45. Safi, M., et al. (2014). Calculation of heat transfer coefficient of MWCNT-TiO_2 nanofluid in plate heat exchanger. *Nanoscience and Nanotechnology, 10*(3), 153–162.
46. Abed, A. M., et al. (2015). Design characteristics of corrugated trapezoidal plate heat exchangers using nanofluids. *Chemical Engineering and Processing—Process Intensification, 87*, 88–103.
47. Khoshvaght-Aliabadi, M., et al. (2016). Thermal–hydraulic performance of wavy plate-fin heat exchanger using passive techniques: Perforations, winglets, and nanofluids. *International Communications in Heat and Mass Transfer, 78*, 231–240.
48. Zheng, D., et al. (2020). Performance analysis of a plate heat exchanger using various nanofluids. *International Journal of Heat and Mass Transfer, 158*, 119993.
49. Fazeli, I., Emami, M. R. S., & Rashidi, A. (2021). Investigation and optimization of the behavior of heat transfer and flow of MWCNT-CuO hybrid nanofluid in a brazed plate heat exchanger using response surface methodology. *International Communications in Heat and Mass Transfer, 122*, 105175.
50. Alklaibi, A., Sundar, L. S., & Mouli, K. V. C. (2022). Experimental investigation on the performance of hybrid Fe_3O_4 coated MWCNT/Water nanofluid as a coolant of a Plate heat exchanger. *International Journal of Thermal Sciences, 171*, 107249.
51. Bahiraei, M., & Monavari, A. (2020). Thermohydraulic characteristics of a micro plate heat exchanger operated with nanofluid considering different nanoparticle shapes. *Applied Thermal Engineering, 179*, 115621.
52. Bhattad, A., Sarkar, J., & Ghosh, P. (2020). Hydrothermal performance of different alumina hybrid nanofluid types in plate heat exchanger. *Journal of Thermal Analysis and Calorimetry, 139*(6), 3777–3787.
53. Attalla, M., & Maghrabie, H. M. (2020). An experimental study on heat transfer and fluid flow of rough plate heat exchanger using Al_2O_3/water nanofluid. *Experimental Heat Transfer, 33*(3), 261–281.
54. Bhattad, A., & Sarkar, J. (2020). Hydrothermal performance of plate heat exchanger with an alumina–graphene hybrid nanofluid: Experimental study. *Journal of the Brazilian Society of Mechanical Sciences and Engineering, 42*(7), 1–10.
55. Kakac, S., Liu, H., & Pramuanjaroenkij, A. (2002) *Heat exchangers: Selection, rating, and thermal design*. CRC Press.

Chapter 3
Double Pipe/Circular Heat Exchanger

Abstract Double pipe heat exchanger is widely used from large to small industries for various applications such as preheating, reheating, digester heating and effluent heating process. Thermal performance of double pipe heat exchanger (DPHE) using nanofluid has been studied in this chapter. Different geometric configuration has been reviewed to analyze their effect on heat transfer rate, effectiveness, NTU and pressure drop. Effect of nanofluid concentration, Reynold number, particle size, flow rate and mass flow rate on thermal performance of DPHE have been analyzed.

3.1 Introduction

With growing population of the world, it was expected that world energy demand will increase by 4.6% in 2021 [1]. To cope up with such rise in energy demand, it is essential to optimize the performance of energy systems. One of the crucial parts of energy systems are the heat exchanger (HE). Heat exchangers play vital role in every industry. Along with shell and tube heat exchanger and plate heat exchangers, double pipe HE (DPHE) is extensively used in oil, food, gas and chemical industries. Double pipe heat exchanger is primarily consisting of two pipes concentric to each other with two fluids (hot and cold fluid) flowing in both pipes. For meeting the high energy demand, design optimization of HE is necessary. Heat transfer in heat exchanger can be enhanced by either optimizing shape of double pipe HE in a way that it does not consume large area and there should not be any pressure drop. Former can be possible by applying different design techniques but late it is almost impossible because increase in pressure drop is inevitable with changing design configuration of any HE.

Yang and Chiang et al. [2] used periodic varying curvature on the inner tube of DPHE and enhanced the Nusselt number by 100% and friction coefficient by 40% compared to smooth tube. Zhang et al. [3] used helical fins as vortex generator on inner tube outer surface of DPHE which allowed heat transfer to increase at the expense of pressure drop. Sharp rise in pressure drop was observed by using small helical pitch. Akpinar et al. [4] used swirl elements at the entrance of inner pipe. Swirl elements helped in enhancing heat transfer rate by 130% compared to smooth

H. M. Ali et al., *Nanofluids for Heat Exchangers*,
https://doi.org/10.1007/978-981-19-3227-4_3

tube. Reducing diameter and raising the number of holes allows heat transfer rate to increase. Addition of swirl caused pressure loss due to formation of drag and immediate contraction. Use of helical wires helped in raising Nusselt number up to 2.64 times and friction factor by 2.74 times compared to smooth tube. Helical wires break down the boundary layer which allow the heat transfer to increase [5]. Choudhari and Taji et al. [6] compared different material of coil wires in DPHE. It was found that wires made of copper had maximum heat transfer rate compared to other materials (stainless steel and aluminum). Friction factor was also raised by decreasing coil pitch. Heat transfer rate was improved by 1.58 times compared to smooth tube. Yildiz et al. [7] used propellers on the inner tube of DPHE to enhance its thermal performance. Heat transfer rate and pressure drop was raised up to 250 and 1000% compared to smooth tube. It was noted that propeller effect at low Reynold number was minimum on heat transfer.

Addition of turbulators in the double pipe heat exchanger also helps in enhancing thermal performance at the expense of pressure drop and friction factor. Zohir et al. [8] used coil wire as a turbulator. Influence of turbulator was to create separation and break boundary layer to enhance heat transfer. Inserting perforated circular rings on the inner tube outer surface helped in raising thermal performance of the system. Thermal performance was raised by increasing number of perforated holes while friction factor was increased by reducing the distance between each circular ring [9].

Apart from modifying the heat transfer area, thermal performance of DPHE can also be enhanced by using thermal fluids with superior properties. Nanofluids has emerged as new and superior thermal fluids for enhancing performance of thermal systems.

3.2 Performance of Nanofluid in Double Pipe Heat Exchanger

Zamzamian et al. [10] investigated performance of double pipe and plate heat exchanger using Al_2O_3 and CuO nanofluid. 37 and 26% heat transfer enhancement was achieved for double pipe heat exchanger using CuO and Al_2O_3 nanofluid while heat transfer for plate heat exchanger was 49 and 38% using CuO and Al_2O_3, respectively. Increased heat transfer was because of chaotic movement of nanoparticles and higher thermal conductivity of nanofluid than water which speed up energy exchange process. Wu et al. [11] studied convective heat transfer and pressure drop in double pipe helical heat exchanger using alumina nanofluid. It was found that effect of nanoparticles on Reynold number was minimal. Also, heat transfer and pressure drop were same for all nanoparticle concentrations used (0.78, 2.18, 3.89, 5.68 and 7.04 wt.%). Goodarzi et al. [12] used nitrogen-doped graphene nanofluid to enhance performance of counter-flow double pipe heat exchanger. Wall temperature reduction was directly related to Reynold number and weight concentration. Also, pressure drop had a direct relation with HE length to its hydraulic diameter, inlet velocity and thermal physical characteristics. 15.87% enhancement of heat transfer coefficient was noticed at 15,000 Re number and 0.06 wt.%. Reddy and Rao et al. [13] compared performance of double pipe heat exchanger with and without helical coil inserts by using TiO_2/EG-H_2O nanofluid. Heat exchanger with helical coils and nanofluid showed better performance than without helical coils because helices enhanced retention time of flow, generated swirl and turbulence in fluid, which raised flow path length. Heat transfer was enhanced to 17.71% at 15,000 Re number and P/d ratio 2.5. It was also noted that friction factor was negligible with helical coils.

Mohammed et al. [14] investigated double pipe with louvered strips (backward and forward arrangement) using different nanofluids (Al_2O_3, CuO, SiO_2, ZnO) to increase heat transfer. Backward louvered strips showed better performance with 1.54 PEC than forward louvered strips with 1.52 at slant angle 30° and 30 mm pitch. It was found that Nusselt number was increased by decreasing and increasing particle size and volume concentration, respectively. Surface area per unit volume was enhanced by decreasing particle size which allowed Nusselt number to increase. Chun et al. [15] found that alumina nanofluid with hydrophobic surface showed better heat transfer than hydrophilic surface alumina. Maximum increase in heat transfer coefficient noticed was 25% for rod shaped alumina nanofluid at 0.5 vol.% and Re number < 200 in comparison with base fluid. Bahmani et al. [16] compared heat transfer between parallel and counter flow double pipe heat exchanger using alumina/water nanofluid. Comparison showed that counter flow performed better than parallel flow double pipe heat exchanger. Also effect of volume fraction was more prominent at higher Reynold number than at lower Reynold number. 30% maximum thermal efficiency and 32.7% enhancement in Nusselt number was achieved for counter flow heat exchanger. Sarfraz and Hormozi et al. [17] investigated effect of biologically prepared Ag/EG-H_2O (50:50) nanofluid. Heat transfer coefficient was greater at higher Reynold number and in turbulent flow because of formation of eddies and local agitation take place. It was suggested that turbulent flow of nanofluid should be used to avoid clogging and aggregation of nanoparticles. Heat transfer coefficient was increased by 67% at 1% volume fraction. Shirvan et al. [18] proposed two-phase mixture model and response surface methodology to analyze heat exchanger effectiveness and heat transfer using Al_2O_3/H_2O nanofluid. Bottom and top walls were supposed to be adiabatic. Heat exchanger effectiveness had a direct relation and Nusselt number had an inverse relation with volume fraction. Nusselt number was enhanced by 57.70% at 0.03 volume fraction and 50–150 Re number. Bahiraei and Hangi et al. [19] applied magnetic field to study working of double pipe counter flow heat exchanger using $Mn_{0.6}Zn_{0.4}Fe_2O_4$/H_2O nanofluid and two-phase Euler–Lagrange method. Results indicated that magnetic field helped in increasing heat transfer due to uniform distribution of nanoparticles. Pressure drop and heat transfer was improved by enhancing magnetic field, volume concentration and particle size.

Moradi et al. [20] did experimentation on double pipe heat exchanger with porous media by using MWCNT/H_2O nanofluid. Results showed that heat transfer had inverse relation with volume fraction and volume flow rate. Also, increasing porous media plates, affected heat transfer rate in a positive way. 35% increase in HTC was achieved at 0.04% mass fraction with 3 porous plates. Arya et al. [21] showed that thermal performance was not affected by fouling when MgO/EG was used as a working fluid. Also, inlet temperature had no significant effect on friction factor and pressure drop. Only 9% HTC was increased at 60 °C inlet temperature. 55.4, 35 and 19% augmentation in thermal conductivity, pressure drop and friction factor was observed at 0.3% weight concentration of MgO/EG nanofluid. Mund et al. [22] prepared hybrid Al_2O_3–TiO_2 nanoparticles by using high energy ball milling technique. It was observed that heat transfer was enhanced with help of hybrid nanofluid Al_2O_3–TiO_2/H_2O due to generation of vortex at inlet. Also, stability observed for hybrid nanofluid was 32 h. Sundar et al. [23] conducted experiment on double pipe heat exchanger with and without wire coil core rod (WCCR) using Fe_3O_4/H_2O nanofluid. Results showed that better heat transfer was achieved by using nanofluid with WCCR than without WCCR because WCCR helped in generating higher turbulent flow. Also, heat transfer was increased by reducing pitch ratio of WCCR. 32.03 and 14.7% enhancement in Nu number was achieved with and without WCCR at 0.06% concentration, 28,994 Re number and 1 pitch ratio, compared to water, respectively (Figs. 3.1 and 3.2).

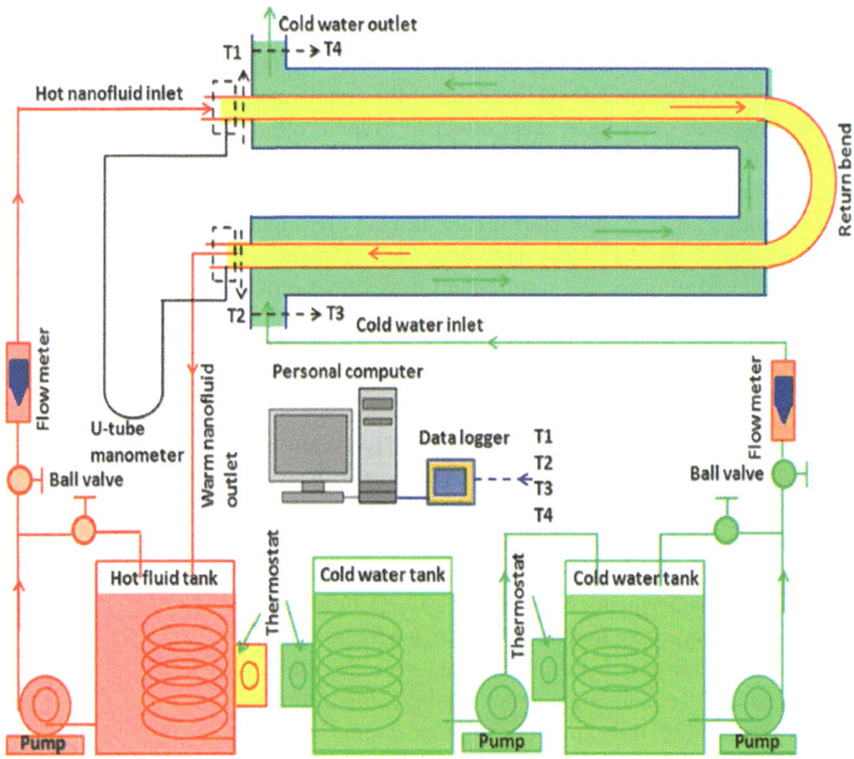

Fig. 3.1 Double pipe with U bend schematic [23]

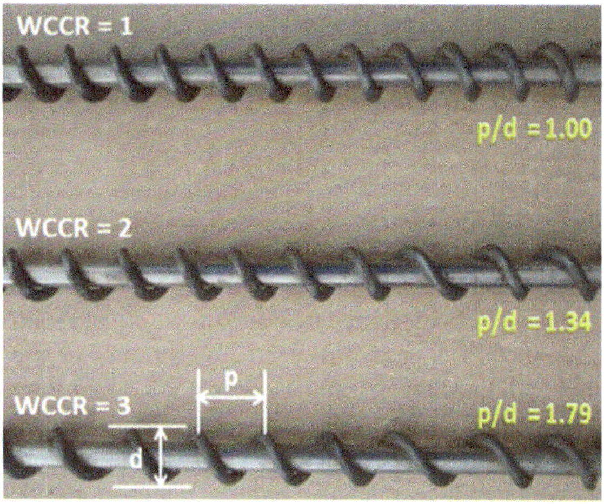

Fig. 3.2 Wire coil with rod inserts [23]

Siavashi and Joibary et al. [24] numerically analyzed performance of double pipe counter flow heat exchanger with porous media. It was suggested that porous medium should be used in inner pipe rather than outer pipe if it had to be used in one pipe (outer or inner) because pressure loss would be higher if porous media was used in outer pipe. Darcy number had a positive impact on heat transfer and PEC. PEC was enhanced by increasing Darcy number and Re number. 4.6 PEC was achieved at 2000 Re number and 0.1 Darcy number.

Shahsavar et al. [25] studied Newtonian and non-Newtonian behavior of hybrid nanofluid (Fe_3O_4-CNT/H_2O) in double pipe heat exchanger. It was concluded that Newtonian hybrid nanofluid had lower heat transfer rate effectiveness and OHTC than non-Newtonian hybrid nanofluid. Viscosity of Newtonian fluid was higher at central region while non-Newtonian fluid had higher viscosity near tube wall. Poongavanam et al. [26] used activated carbon nanoparticle with solar glycol base fluid in shot peened double pipe HE. Stanton number was enhanced by using shot peened tube by 28%. Stanton number was reduced by increasing flow rate or Reynold number. Nu number and CHTC was raised by 57.06 and 71% at 0.6% volume concentration and 0.04 kg/s flow rate. With increment of CHTC at 0.6% concentration, friction factor was also 3.47 times greater than base fluid. Poongavanam et al. [27] applied shot peening surface modification technique in double pipe heat exchanger. Reynold number was reduced by enhancing volume concentration of MWCNT in solar glycol at given flow rate. Shot peened tube enhanced Nu number by 28.2%. Friction factor was reduced by raising flow rate. CHTC was inversely proportional to Reynold number. Pressure drop of nanofluid was 1.56 times more than base fluid at 0.6% volume concentration and 0.08 kg/s flow rate. Gnanavel et al. [28] enhanced thermal performance of double pipe heat exchanger using twisted tapes and different nanofluids (TiO_2, BeO, CuO, ZnO). TPF was reduced by raising velocity of fluid. Use of twisted tape helped in mixing and resisting of flow to enhance TPF. 1.55, 1.53, 1.51, and 1.49 thermal performance factor was achieved for TiO_2, BeO, ZnO and

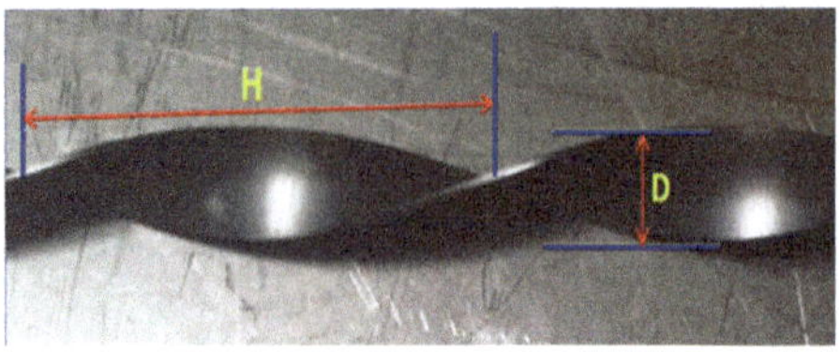

Fig. 3.3 Twisted tape insert [29]

CuO, respectively. Kumar et al. [29] enhanced Nusselt number by 38.75 and 14.76% with and without twisted tape at 0.06% concentration (Fe_3O_4/H_2O), twist ratio of 10 and 30,000 Re number respectively. Nu number was enhanced by decreasing twist ratio and enhancing volume concentration and Reynold number. Hosseinian et al. [30] studied effect of vibration on heat transfer performance of MWCNT/water in flexible double pipe heat exchanger. Vibration not only helped in enhancing heat transfer coefficient but also reduced nanoparticles deposition. Application of vibration was more useful in terms of lower flow rates. 100% increment in heat transfer coefficient was observed at minimum mass fraction 0.04% and maximum vibration level 9 m/s^2 (Fig. 3.3).

Maddah et al. [31] prepared hybrid Al_2O_3–TiO_2/H_2O nanofluid without any surfactant. Twisted tapes were inserted in double pipe heat exchanger. Exergy efficiency was boosted due to increment in Re number, volume concentration and reduction of twist ratio. It was noted that twisted tape had more positive effect on heat transfer rate than nanofluid. But their combined use was more fruitful in reducing exergy destruction. Eiamsa-ard et al. [32] applied double twisted tape in double pipe heat exchanger using TiO_2/H_2O nanofluid. Heat transfer, thermal performance and friction loss had an inverse relation with twist ratio. Nusselt number was also increased at lower twist ratio because twisted tape allowed enhancement in swirling which resulted in thinning of thermal boundary layer. TPF of TiO_2/H_2O nanofluid was greater than base fluid by 25.9, 41.1 and 56.6% at 2.5, 2 and 1.5 twist ratio, respectively. Sarafraz et al. [33] enhanced stability of CNT/H_2O nanofluid by functionalizing it with COOH group which helped in increasing thermal conductivity of nanofluid due to reduced deposition and agglomeration. OHTC was raised by raising inlet temperature of working fluid. Thermal conductivity and thermal performance were enhanced up to 56 and 44% at 0.3% weight concentration of CNT/H_2O nanofluid. Little drawback noticed was increment in pressure drop and friction factor up to 11 and 9%. Mousavi et al. [34] numerically showed that heat transfer was enhanced by increasing magnetic field in sinusoidal double pipe heat exchanger using Fe_3O_4/H_2O nanofluid. Nu number was increased to 25% by raising shape coefficient of sinusoidal HE from 0.2 to 0.3. Figure 3.4 represents sinusoidal double pipe heat exchanger.

Kumar et al. [35] experimentally analyzed performance of Fe_3O_4/water nanofluid in double pipe heat exchanger with and without longitudinal strips. Nanofluid performed better with longitudinal strips at lower aspect ratio than no inserts in terms of Nu number and heat transfer enhancement. 41.29 and 14.7% enhancement in Nu number was achieved at 28,954 Re number, 0.06% volume concentration with and

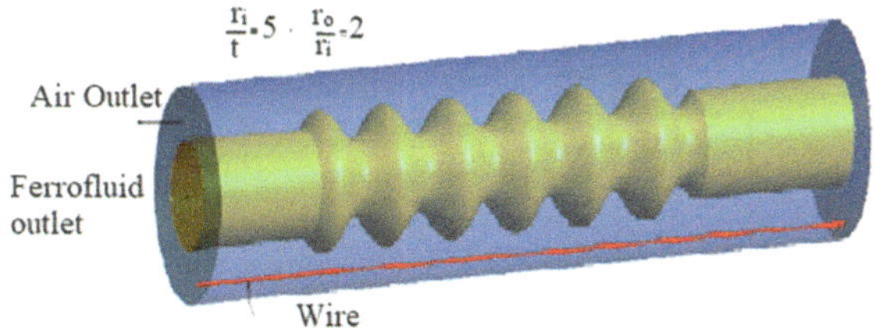

Fig. 3.4 Sinusoidal double pipe heat exchanger [34]

without longitudinal strips (AR = 1), respectively Saeedan et al. [36] did CFD analysis and neutral network modeling using different nanofluid (Cu, CuO and CNT) in double pipe with helical baffle (helix angle 40) heat exchanger. CFD analysis showed that Nu number for Cu/water and CuO/water was enhanced by enhancing volume concentration but it was inversely related with CNT/water concentration. Raei et al. [37] prepared alumina water nanofluid without using any surfactant. Results revealed that there was no sensible enhancement in heat transfer by raising volume concentration. Also, HTC was reduced by raising flow rate of nanofluid. 23 and 25% enhancement in heat transfer coefficient and friction factor was observed at 0.15% volume concentration, respectively (Fig. 3.5).

Hussein et al. [38] used aluminum nitride/EG hybrid nanofluid in double pipe heat exchanger. Effective viscosity was reduced by raising temperature because of unstable structural combination. Nu number and friction factor was enhanced by 35 and 12.5% using AlN/EG nanofluid. AlN was highly suggested for industrial applications. Schematic of circular tube is given in Fig. 3.6.

Shakiba and Vahedi et al. [39] investigated numerically effect of non-uniform magnetic field on ferro nanofluid in double pipe heat exchanger. Force produced with help of magnetic field caused formation of vortices which moved ferro fluid toward internal pipe, this allowed diffusion of cold boundary layer and warm ferro fluid boundary layer, which resulted in enhancing Nusselt number. Friction factor and pressure drop were also raised by applying non-uniform magnetic field. Optimum magnetic field found was between 1.33×10^6 and 2.37×10^6. Doruk et al. [40] found

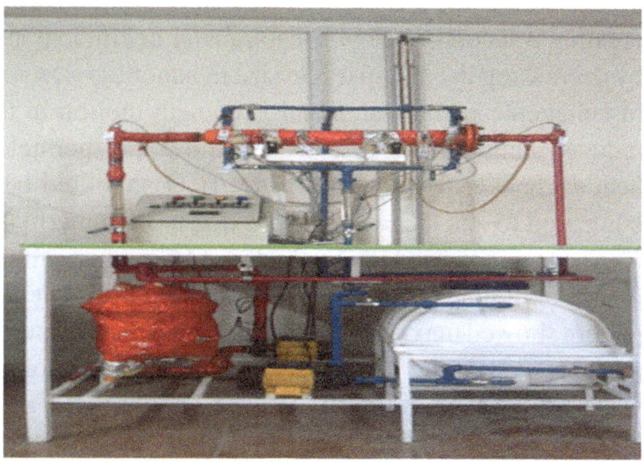

Fig. 3.5 Double pipe heat exchanger experimental setup [37]

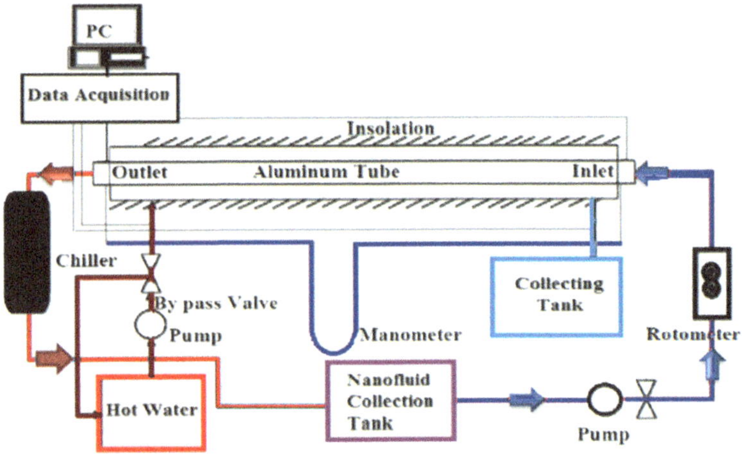

Fig. 3.6 Double pipe schematic [38]

that there was zero improvement in heat transfer coefficient when n-nonadecane was used at 0.42 and 0.84% volume fraction while about 10% enhancement in HTC was attained at 1.68% volume fraction. Similarly, maximum improvement ratio was shown by n-nonadecane at 1.68% volume fraction which was equal to 1.05–1.09. Aghayari et al. [41] experimentally studied Al_2O_3/H_2O nanofluid in mini double pipe exchanger. 1.71% higher thermal efficiency was achieved at 0.1% concentration of nanofluid as compared to base fluid. Thermal boundary layer thickness was reduced due to motion of nanoparticles near wall and reduction of viscosity at wall region. Ghasemi et al. [42] applied twisted tape in double pipe heat exchanger to enhance heat transfer performance using MgO/water nanofluid. Results showed that pressure drop was raised from 2.5 to 2.8747 kPa at 1% volume concentration and twist ratio of 4.4. Moreover, 3122.08 W/m^2K heat transfer coefficient was achieved at Re = 28,000, concentration = 1% and twist ratio = 2.6.

El-Behery et al. [43] enhanced heat transfer coefficient, effectiveness of double pipe heat exchanger by adding swirl effect which induced turbulent effect resulting in creating a secondary flow. Heat transfer coefficient and effectiveness was enhanced by decreasing the nanoparticle size in nanofluid. Decreasing nanoparticle size helped in enhancing surface area and Brownian motion in the nanofluid enhanced due to reduced nanoparticle size which allow the thermal performance of double pipe heat exchanger to enhance. Al_2O_3/H_2O nanofluid had better thermal performance compared to ZnO, CuO and TiO_2 due to high thermal conductivity and low density of Al_2O_3 nanoparticles. Asadi et al. [44] numerically compared the thermal performance of double pipe heat exchanger with different pipe shapes. Pipe with spherical pattern showed maximum enhancement in thermal performance compared to smooth, corrugated and tube with axial fins. Also, comparison between two hybrid nanofluids was made, Ag–MoS_2 was found to be better than Fe_3O_4–SiO_2 with enhancement in heat transfer coefficient by 62.21% at 1 vol.% concentration (Fig. 3.7).

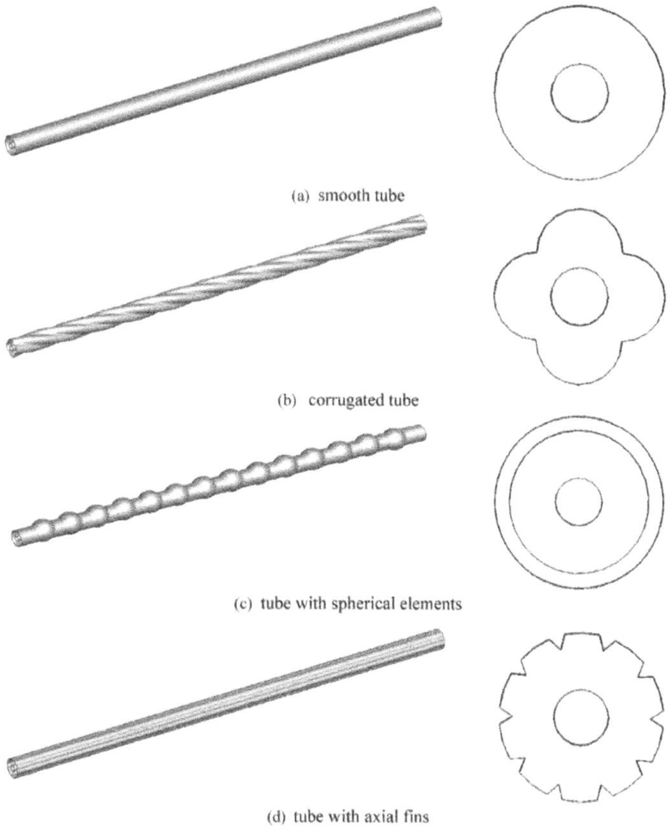

Fig. 3.7 Different geometries of tube used by [44]

Kerur et al. [45] studied effect of number of turns in helical double tube heat exchanger using different nanofluids. Increasing number of turns helped in enhancing thermal performance of the system in terms of effectiveness, LMTD, overall heat transfer coefficient and heat transfer rate due to increased heat transfer surface area and turbulence effect (Fig. 3.8).

Noorbaksh et al. [46] numerically evaluated performance of double pipe heat exchanger with twisted tape using three different water based nanofluids (CuO, Al_2O_3 and SiO_2). CuO/H_2O had better thermal performance due to high thermal conductivity of CuO nanoparticles. Also, thermal performance of the system was raised due to insertion of twisted tape on the outer tube which helped in creating more intense turbulent effect resulting in increased heat transfer (Fig. 3.9).

Karouei et al. [47] analyzed the hybrid nanofluid (Ag-Graphene/H_2O and MWCNT-Fe_2O_3/H_2O) effect on thermal performance of double pipe heat exchanger inserted with 12 blades turbulator as shown in Fig. 3.10.

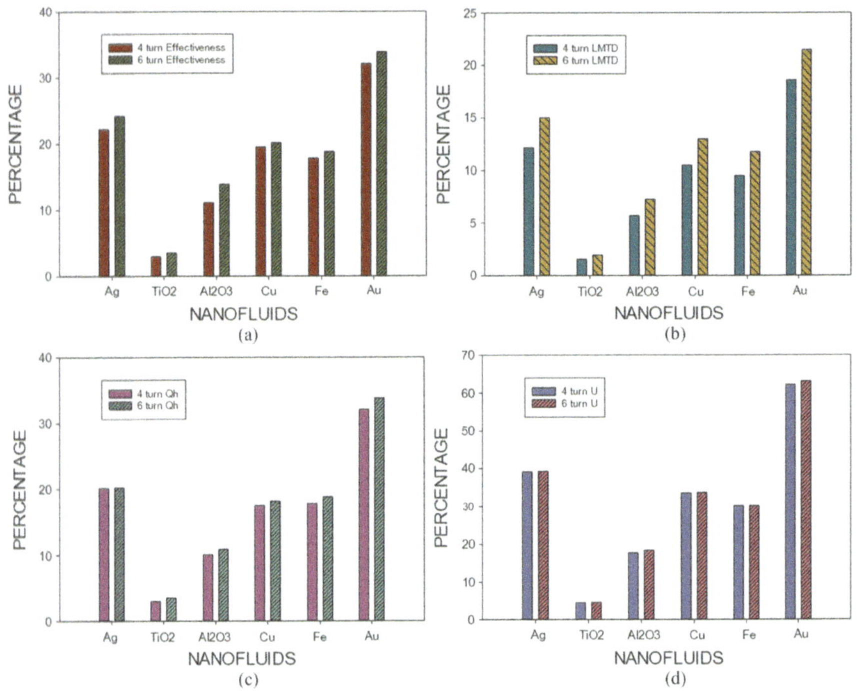

Fig. 3.8 Effect of number of turns on **a** Effectiveness, **b** LMTD, **c** Q_h, **d** U using different nanofluids [45]

Use of turbulator creates a secondary flow with swirls in laminar regime helping in raising heat transfer between the hot and cold fluids. Ag-Graphene nanofluid had better thermal performance than MWCNT-Fe_2O_3 as shown in Fig. 3.11. Thermal performance was reduced with increased mass flow rate for both nanofluids.

Shahsavar et al. [48] showed that performance of double pipe heat exchanger was increased with increase in twist pitch up to an optimum point using non-Newtonian nanofluids. Increasing twist ratio generates more strong vortices which allows heat transfer to increase with increase in pressure drop. Up to optimum point, increment in heat transfer dominates pressure drop but above optimum twist pitch pressure drop dominates which cause the reduction in PEC as shown in Fig. 3.12.

Bashtani et al. [49] numerically investigated the effect of gear disc as turbulator in double pipe heat exchanger using Al_2O_3/H_2O nanofluid. Heat transfer was enhanced by using gear turbulators due to breaking of boundary layer and proper mixing fluid allowing heat transfer to increase between the fluid layers. It was also found that heat transfer coefficient achieved at low Reynold number using Al_2O_3/H_2O nanofluid along with disc turbulators was same as achieved at high Reynold number without

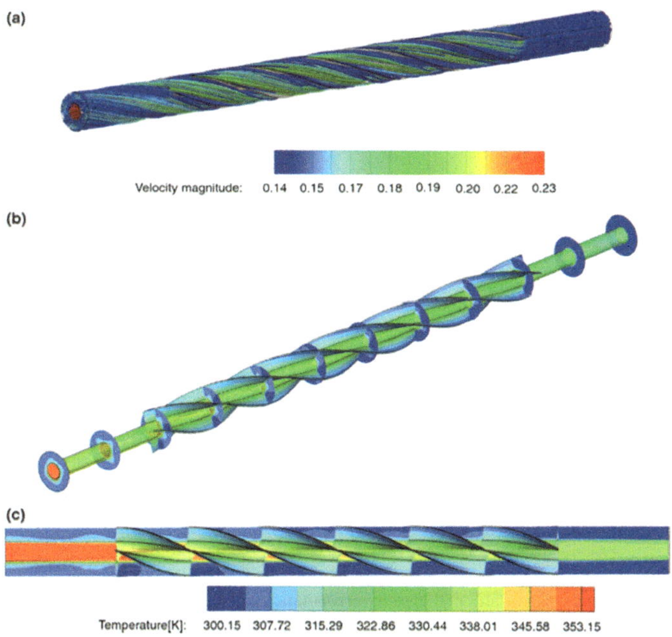

Fig. 3.9 **a** 3D streamline with contour of velocity magnitude. **b** 3D contours of temperature. **c** s2D contours of temperature at $Y = 0$ mm and Re = 5492 for CuO/H_2O nanofluid [46]

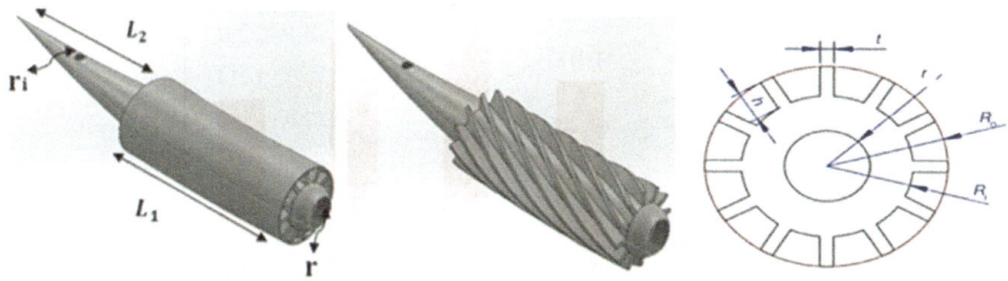

Fig. 3.10 Schematic of curved turbulators [47]

turbulators. SST$k - \omega$ model was used to numerically optimize the performance of double pipe heat exchanger which is given below:

Turbulence Kinetic Energy

$$U_j\left(\frac{\partial \rho k}{\partial x_j}\right) = \tau_{ij} - \beta^* k\rho\omega + \frac{\partial}{\partial x_j}\left[(\mu + \sigma_k\mu_t)\frac{\partial k}{\partial x_j}\right] \tag{3.1}$$

Specific Dissipation Rate

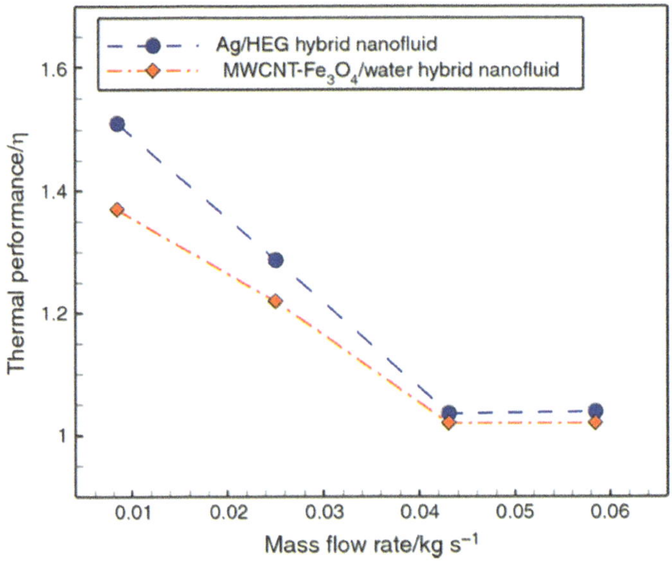

Fig. 3.11 Effect of mass flow rate on thermal performance of heat exchanger using hybrid nanofluid [47]

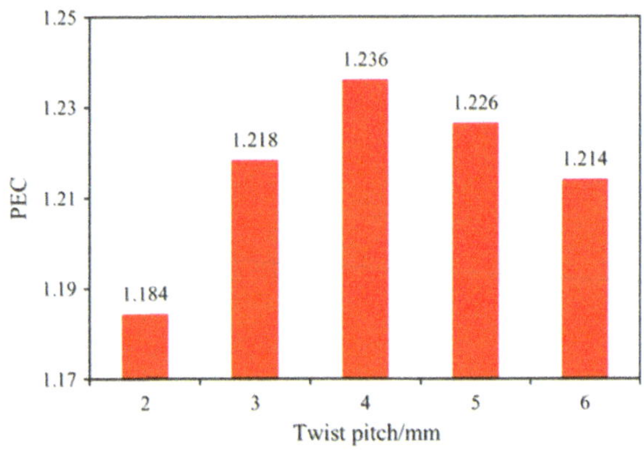

Fig. 3.12 Effect of twist pitch on performance evaluation criteria

$$U_j\left(\frac{\partial\rho\omega}{\partial x_j}\right)=\frac{\gamma}{V_t}\tau_{ij}\frac{\partial u_i}{\partial x_j}-\beta\omega^2+\frac{\partial}{\partial x_j}\left[(\mu+\sigma_k\mu_t)\frac{\partial\omega}{\partial x_j}\right]+2(1-F_1)\rho\sigma_{\omega 2}\frac{1}{\omega}\frac{\partial k}{\partial x_j}\frac{\partial\omega}{\partial x_j} \quad (3.2)$$

Values for constant for above equations are:

$$\sigma_{k1} = 0.85, \sigma_{\omega 1} = 0.5, \beta_1 = 0.0750,$$
$$\beta^* = 0.09, k = 0.41, \gamma_1 = \frac{\beta_1}{\beta^*} - \frac{\sigma_{\omega 1}.k^2}{\sqrt{\beta^*}}, a_1 = 0.31$$

and

$$F_1 = \tanh\left(\arg_1^4\right) \tag{3.3}$$

$$\arg_1 = \min\left[\max\left(\frac{\sqrt{k}}{0.09\omega y}\right); \frac{500\upsilon}{y^2\omega}; \frac{4\rho\sigma_{\omega 2}k}{\mathrm{CD}_{k\omega}y^2}\right] \tag{3.4}$$

$$\mathrm{CD}_{k\omega} = \max(2\rho\sigma_{\omega 2}\left(\frac{1}{\omega}\right)\frac{\partial k}{\partial x_j}\frac{\partial \omega}{\partial x_j}, 10^{-20} \tag{3.5}$$

Eddy viscosity can be evaluated using:

$$V_t = \frac{a_1 k}{\max(a_1\omega, \varnothing F_2)} \tag{3.5}$$

$$F_2 = \tanh\left(\arg_2^2\right) \tag{3.6}$$

$$\arg_2 = \max\left[\frac{2\sqrt{k}}{0.09\omega y}; \frac{500\upsilon}{y^2\omega}\right] \tag{3.7}$$

Dalkilic et al. [50] optimized performance of double pipe heat exchanger using fins and nanofluids. Addition of fins in double pipe heat exchanger helped in reducing the total number of tubes as double pipe fins create more heat transfer surface area compared to bare double pipe. Fins also help in reducing the pressure drop due to reduced number of tubes. It was recommended to use nanofluid with high thermal conductivity nanoparticles such as MWCNT and graphene. Only disadvantage of using fins and nanofluids in double pipe heat exchanger are enhanced fouling and lower cleanliness factor. Cleanliness factor is given as:

$$F = \frac{U_{\mathrm{of}}}{U_{\mathrm{oc}}} \tag{3.8}$$

If CF = 1, it shows a clean case and if CF < 1, it shows fouled case.

Thermal Analysis of Double Pipe Heat Exchanger

Inner Tube

Energy balance equation between hot and cold fluid is given by [50]:

$$\dot{Q} = m_1 C_{p,1}(T_{1\,\text{out}} - T_{1\,\text{in}}) = m_2 C_{p2}(T_{2\,\text{in}} - T_{2\,\text{out}}) \tag{3.9}$$

Subscript 1 indicates the fluid in the inner tube and subscript 2 shows the fluid flowing in the outer tube.

Number of parallel pipes in DPHE can be calculated by using the following [50]:

$$n_p = \frac{\dot{m}_1}{\frac{\pi d_{1i}^2}{4} w_1 \rho_1} \tag{3.10}$$

where w is the flow velocity of fluid flowing on the inner tube.

Annulus Side

Wetted perimeter for the pressure drop (P_w), cross sectional area (A_c), hydraulic diameter for the pressure drop ($d_{2,h}$), and fluid velocity of the annulus side (w_2) are calculated by using the following relations [50]:

$$P_w = \pi\left(d_{2,i} - d_{1,o}N_t\right) + 2H_f N_f N_t \tag{3.11}$$

$$A_c = \frac{\pi}{4}\left(d_{2,i}^2 - d_{1,o}N_t\right) - \delta H_f N_f N_t \tag{3.12}$$

$$d_{2,h} = \frac{4A_c}{P_w} \tag{3.13}$$

$$w_2 = \frac{\dot{m}_2}{A_c n_p \rho_2} \tag{3.14}$$

Considering flow to be laminar on the annulus side, following relation can be used for evaluating the Nusselt number [50]:

$$\text{Nu}_2 = 1.86\left(\text{Re}_1 \underset{1}{\text{Pr}}\left(\frac{d_{2,h}}{L}\right)\right)^{\frac{1}{3}} \left(\frac{\mu_2}{\mu_w}\right)^{0.14} \tag{3.15}$$

μ_w shows the viscosity evaluated at wall temperature T_w which can be calculated by using the following relations:

$$T_w \approx \frac{1}{2}\left(\frac{T_{2,i} + T_{2,o}}{2} + \frac{T_{1,i} + T_{1,o}}{2}\right) \tag{3.16}$$

$$\frac{\mu}{\mu_w} = \left(\frac{T}{T_w}\right)^{-1.6} \tag{3.17}$$

Overall Heat Transfer Coefficient

Overall heat transfer coefficient for clean (U_{oc}) and fouled (U_{of}) heat exchanger is given as:

$$U_{oc} = \frac{1}{\frac{A_t}{A_i h_1} + \frac{A_t \ln\left(\frac{d_{1,o}}{d_{1,i}}\right)}{2\pi k 2L} + \frac{1}{\eta_0 h_2}} \tag{3.18}$$

$$U_{of} = \frac{1}{\left(\frac{A_t}{A_i h_1}\right) + \left(\frac{A_t}{A_i}\right) R_{f_i} + \frac{A_t \ln\left(\frac{d_{1,o}}{d_{1,i}}\right)}{2\pi k 2L} + \frac{R_{fo}}{\eta_0} + \frac{1}{\eta_0 h_2}} \tag{3.19}$$

R_{fi} and R_{fo}(m^2 W^{-1} °C^{-1}) indicates fouling resistance for tube and annulus side.

3.3 Performance of Nanofluids in Circular Tube

Heris et al. [51] used Al_2O_3/H_2O nanofluid in circular tube to analyze heat transfer performance. It was concluded that thermal conductivity was not the only reason for heat transfer enhancement for nanofluid, it was also enhanced due to nanoparticles Brownian motion, chaotic movement and migration of nanoparticles. Peclet number and volume concentration had a direct relation with HTC. Fotukian and Esfahany et al. [52] compared experimental results with Maiga et al. [53] correlation and showed that Maiga et al. [53] overestimated experimental result of convective heat transfer of nanofluid. Convective heat transfer coefficient ratio was inversely related to Reynold number. Also, heat transfer was not much effected by increasing volume concentration. 25 and 20% increment in heat transfer coefficient and pressure drop was observed at 0.3% volume concentration compared to water, respectively. Bianco et al. [54] numerically investigated turbulent effect on heat transfer performance in circular tube heat exchanger using Al_2O_3/H_2O nanofluid and finite volume method. Numerical results were closed to the simulation results obtained in [55–58]. Heat transfer was enhanced with Re number and particle concentration. Sajadi and Kazemi et al. [59] investigated pressure drop and CHTC in circular tube using TiO_2/H_2O nanofluid. Results of experiment were underestimated when compared to Pak et al. [60] and Xuan et al. [61] Nu number correlations. It was also found that heat transfer enhancement was not affected by enhancing volume concentration, so a new correlation was proposed which was independent of particle concentration, given in Table A.2. 22 and 25% enhancement in heat transfer coefficient and pressure drop was noticed at Re = 5000 and 0.25% concentration.

Fotukian and Esfahany et al. [62] experimentally investigated heat transfer in circular tube using γ-Al_2O_3/H_2O. Augmentation in heat transfer was achieved due to expulsion of nanoparticles from wall to bulk of flow. Maximum increment in heat transfer coefficient was 48% at 0.054% concentration and 10,000 Re number. 40% increased pressure drop was also achieved as a drawback at 0.135% concentration and Re = 20,000. Fard et al. [63] numerically investigated performance of circular tube in presence of Cu/H_2O nanofluid using single-phase and two-phase model. It was found that two-phase model was more precise with 11.08% average relative error than single-phase model with 17.6% average relative error at 2% volume concentration. Heat transfer coefficient ratio was enhanced more at higher Peclet number than at lower Pe number. 27.8% increment in heat transfer coefficient was achieved at 3% volume concentration. Sundar et al. [64] investigated effect of twisted tape (twist ratio 0 to 83) in circular tube with Al_2O_3/H_2O as working fluid. Heat transfer coefficient was enhanced by 30.30 and 42.17% at 22,000 Re number and 0.5% concentration for plain and circular tube with twisted tape inserts (twist ratio 5) in comparison with water, respectively. New correlation for Nu number and friction factor was developed given in Table A.2. Bianco et al. [65] numerically investigated effect of nanofluid (Al_2O_3/H_2O) in circular tube using single and two-phase models with either constant or temperature dependent properties. Heat transfer was enhanced by enhancing concentration and wall shear stress. Heris et al. [66] numerically studied effect of nanoparticle size and concentration on Nu number in circular tube under laminar flow. Result showed that Nu number was reduced by raising the Al_2O_3 nanoparticle size and decreasing concentration. Moreover, local Nu number was inversely related to thermal boundary layer thickness. Sharma et al. [67] used Al_2O_3/H_2O nanofluid in circular tube under transition flow. Experimental results showed deviation from Gnielinski equation [68] in single-phase transitional flow fluid. Heat transfer rate was found to be 47.71 and 36.96% at 0.1% volume concentration and Re = 9000 and 3000 with twist ratio of 5 in comparison with plain tube, respectively. New correlation was proposed for Nu number and friction factor, given in Table A.2.

Hojjat et al. [69] proposed new correlation for Nu number of non-Newtonian nanofluid which was dependent on Re number and Pr number. By changing Peclet number from 210,000 to 360,000, HTC was increased to 71, 68 and 67% for CuO/CMC, γ-Al_2O_3/CMC and TiO_2/CMC, respectively. Heris et al. [70] analyzed convective heat transfer in circular tube using Cu/H_2O nanofluid. Result demonstrated that enhancement in HTC achieved by experimental result was greater than HTC obtained from single-phase heat transfer correlation. HTC was higher at higher Pe number as compared to smaller Pe number because of variation in flow structure due to Cu nanoparticles interaction and fluctuation which helped in enhancement in heat transfer. HTCR was enhanced to 1.45 at 2% volume concentration and Pe = 6700. Saeedinia et al. [71] found that Hamilton-Crosser [72] and Yu and Choi [73] models underestimated nanofluid effective thermal conductivity. Thermal conductivity was increased and specific heat was decreased by enhancing volume concentration of nanofluid. Enhancement of 6.2% in thermal conductivity was achieved at 2% mass fraction. Suresh et al. [74] found enhancement in thermal conductivity and viscosity of hybrid nanofluid near centerline because of nanoparticle motion toward tube centerline due to thermophoresis and Brownian motion. Temperature difference between fluid bulk temperature and wall temperature was found to be lower for hybrid nanofluid Al_2O_3–Cu/H_2O as compared to Al_2O_3/H_2O. Average increment of 10.94% in Nu number was achieved at 0.1% hybrid nanofluid concentration. Rashidi et al. [75] studied entropy formation and compared single-phase and two-phase modeling approaches. Modeling approaches used in this study were volume of fluid (VOF), Mixture and Eulerian Models. Maximum and minimum total entropy generation was found for Eulerian approach and single phase with increasing Re number, respectively. Kim et al. [76] used two different nanofluids (Al_2O_3/H_2O and Amorphous Carbonic/H_2O) in circular tube under laminar and turbulent flow. Al_2O_3/water nanofluid performed better in comparison with A'C'/H_2O nanofluid. 15 and 20% enhancement in CHTC was observed for Al_2O_3/H_2O nanofluid under laminar and turbulent flow, respectively, while no heat transfer coefficient enhancement was observed for amorphous carbonic nanofluid under turbulent flow and only 8% enhancement in laminar flow.

Bianco et al. [77] numerically studied performance of Al_2O_3/H_2O nanofluid in circular tube using mixture model approach at constant wall temperature. Nu number was increased by increasing concentration with increment in pressure drop and entropy generation. Two criteria (PEC and entropy generation) were used in order to evaluate system performance. Also, it was suggested that base fluid (H_2O) was a better choice for maximizing energy efficiency. Eiamsa-ard et al. [78] enhanced heat transfer in heat exchanger tube using overlapped dual twisted tape and TiO_2/H_2O nanofluid. Twist ratio used in this study were 1.5, 2.0 and 2.5. Result showed that thermal performance, Nu number and friction factor were enhanced by increasing concentration and decreasing twist ratio. Thermal performance was 4.2% greater than water at 0.21% TiO_2 concentration. Also, thermal performance was increased at lower Re number. New correlation for Nu number, friction factor and thermal performance were developed. Tahir and Mital et al. [79] numerically analyzed heat transfer in circular tube using Euler-Lagrangian approach. Effect of Re number, particle size and volume concentration were studied using Al_2O_3/H_2O nanofluid. Re number had greater impact on HTC as compared to volume concentration. Re number and volume concentration had a direct linear relation with HTC while HTC showed non-linear parabolic decrement with increase in particle size. Heyhat et al. [80] experimentally investigated effect of Al_2O_3/H_2O nanofluid under laminar flow in circular tube. Results showed that single-phase correlation failed to predict heat transfer coefficient enhancement. New correlation for thermal conductivity and dynamic viscosity were proposed in this study. 32% enhancement in HTC was observed at 2% volume concentration in comparison with water. Sundar et al. [81] studied heat transfer and friction factor in circular tube with twisted tapes insert under laminar flow and Al_2O_3/H_2O nanofluid. New correlation for Nu number and friction factor was developed dependent on twist ratio, Re and Pr number, given in Table A.2. No significant enhancement in pressure drop was observed in this study. Friction factor for twisted tape inserts was 1.0652 and 1.0413 times greater than water and plain tube at $Re = 2200$ and 0.5% volume concentration, respectively.

Zanjani et al. [82] prepared graphene/water nanofluid by using PVA as a surfactant. Nanofluid thermal conductivity was raised to 10.3% at 45 °C temperature and 0.02% concentration. It was concluded that increment in volume concentration was more useful at low Reynold number than at high Re number. Moreover, viscosity of nanofluid was enhanced by raising concentration and reducing temperature. Maximum increment of 14.2% in HTC was noted at Re = 1850 and 0.02% concentration. Hojjat et al. [83] showed that Peclet number had no notable effect on relative HTC. On the other hand, HTC was reduced by increasing axial distance (inlet to outlet of tube) because of enhanced thermal boundary layer thickness at certain concentration and Peclet number. At Pe = 140,000, HTC was enhanced from 7.9 to 21.2, 6.7 to 18.4 and 7.4 to 21.3 for CuO/CMC, TiO_2/CMC and γ-Al_2O_3/CMC, when volume concentration was raised from 0.1 to 1.5%, respectively. Hojjat et al. [84] prepared non-Newtonian nanofluid γ-Al_2O_3, TiO_2, CuO with aqueous carboxymethyl cellulose solution as base fluid. Experimental data and single-phase non-Newtonian fluid correlation were not in good agreement. A new empirical correlation was proposed for Nu number which was dependent on Re number and Pr number. TiO_2/CMC had more increment in HTC than other two nanofluids used in this study. Bianco et al. [85] studied effect of volume concentration and Re number on thermal and frictional entropy generation in circular tube under turbulent flow. Thermal entropy generation was reduced and frictional entropy generation was enhanced by raising concentration of nanoparticles because by increasing concentration Nu number was increased with increment in viscosity which in turn reduced thermal dissipation. On the other hand, Re number was directly and inversely related to frictional entropy and thermal entropy generation, respectively. Results showed that it was necessary to use low concentration of nanoparticle to diminish total entropy generation which was minimized at constant velocity. Khairul et al. [86] analyzed performance of three nanofluids (CuO, Al_2O_3, ZnO) in helical coiled heat exchanger. Generation of thermal boundary layer was disturbed by chaotic movement of nanoparticles in base fluid. 7.14 and 6.14% enhancement and reduction in heat transfer coefficient and entropy generation was observed, respectively. Entropy generation was inversely related to volume fraction and flow rate. Mohammed and Narrien et al. [87] numerically studied effect of geometrical parameters on performance of helical coiled tube heat exchanger using nanofluid. Nusselt number and heat transfer was enhanced by reducing annulus diameter and enhancing diameter of inner tube under laminar flow. Pressure drop was not affected by helix pitch and annulus diameter. But pressure drop was reduced due to increase in helix radius. 29 and 34% enhancement in performance index and effectiveness was achieved for counter flow helical coiled heat exchanger, better than parallel flow heat exchanger.

3.4 Discussion

Comparison of Nusselt number is shown in Fig. 3.13 for different nanofluids and at optimum Reynold number and concentration. Maximum Nusselt number was achieved for SiO_2/H_2O followed by TiO_2/H_2O, BeO/H_2O and ZnO/H_2O. Maximum Nu number of 365 was achieved at Reynold number = 42,000 and 4% volume concentration for SiO_2/H_2O in double pipe heat exchanger. 346, 345 and 340 Nu number was attained for TiO_2/H_2O, BeO/H_2O and ZnO/H_2O at Re = 10,000, respectively. From Fig. 3.13, minimum Nusselt number was gained by using $MWCNT/H_2O$, it was due to the low Reynold number as compared to other nanofluids. Nusselt number gain is enhanced at higher Reynold number because higher Reynold means turbulent flow regime. Hence, in turbulent conditions there is better collision and mixing of nanoparticles which enhanced Brownian motion of particles inside fluid resulted in enhancement of Nu number as well as frictional losses (Fig. 3.14).

Figure 3.15 represent relation of friction factor for Reynold number. Maximum friction factor obtained was 0.25 for TiO_2/H_2O. This highest friction factor achieved was due to twisted tapes which caused enhancement in frictional losses. While for plain tube 80% less friction was achieved as compared to tube with twisted tapes. Minimum friction factor obtained was for MgO/EG at Re = 10,000 and 0.2 wt. % as compared to other nanofluids shown in Fig. 3.15. Summary of above studied literature is given in Table 3.1.

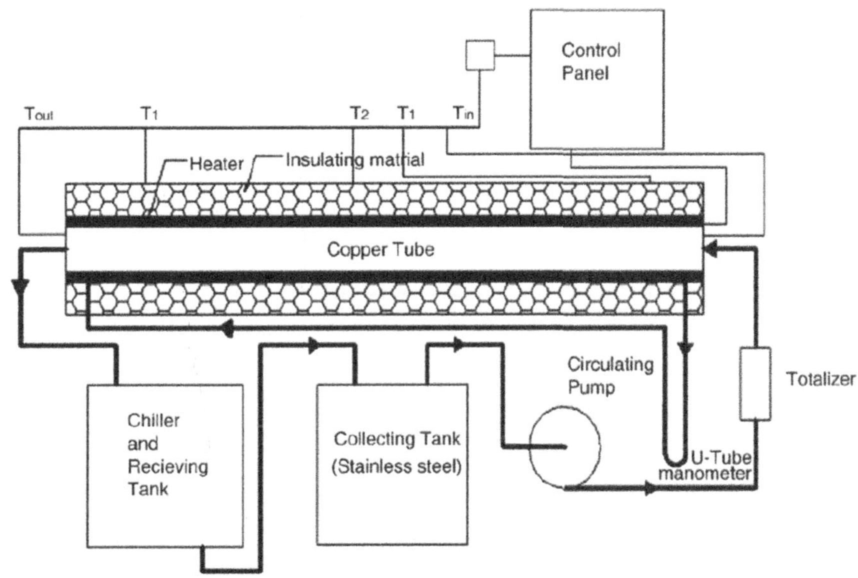

Fig. 3.13 Schematic of circular tube heat exchanger [67]

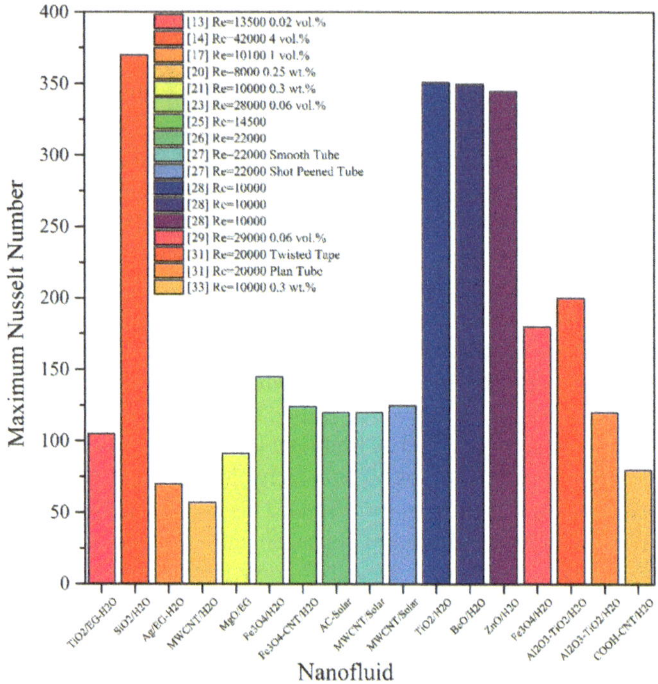

Fig. 3.14 Nusselt number variation for different nanofluids

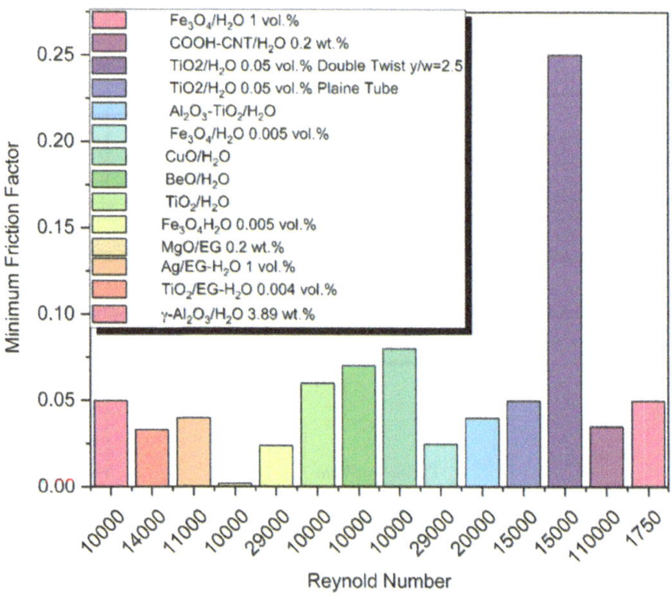

Fig. 3.15 Friction factor versus Reynold number for various nanofluids

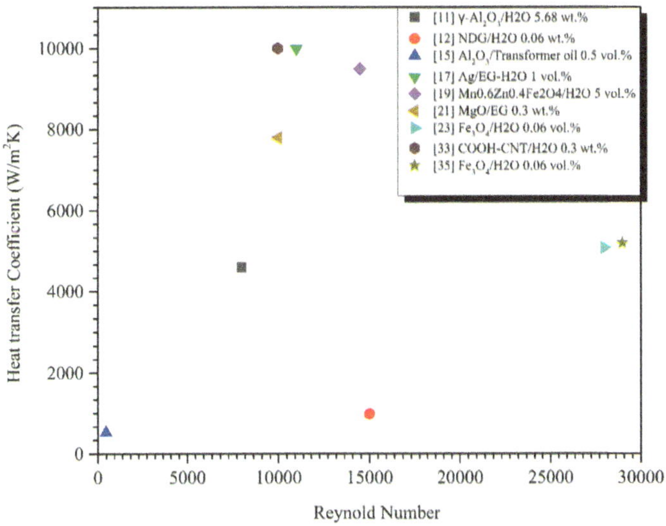

Fig. 3.16 Maximum heat transfer coefficient as a function of Reynold number

Figure 3.16 represents effect of Reynold number on heat transfer coefficient for different nanofluids in double pipe heat exchanger. Heat transfer coefficient depends on Reynold number and thermo physical properties of nanoparticles. CNT/H_2O showed maximum heat transfer coefficient than other nanofluids shown in Fig. 3.15 because thermal conductivity of CNT was highest among all other nanoparticles but less than graphene. Due to better thermal properties of CNT, heat transfer was maximum at Reynold number = 10,000. Heat transfer also enhanced due to enhancement in Re number because higher Re (turbulent flow) caused reduction in agglomeration of nanoparticles and thermal boundary layer thickness thus, resulting in the enhancement of heat transfer coefficient and heat transfer rate.

Table 3.1 Summary of literature on the usage of nanofluid in double pipe heat exchanger

References	Heat exchanger	Nanofluid	Nanoparticle size	Surfactant	Concentration	Flow	Outcomes
[10]	Double pipe and plate heat exchanger	Al_2O_3, CuO/EG	20 nm	SDS for CuO/EG	0.1–1 wt.%	Turbulent	49 and 37% heat transfer were achieved by using CuO/EG in plate and double pipe heat exchanger, respectively
[11]	Double pipe helical	γ-Al_2O_3/H_2O	40 nm	No surfactant used	0.78–7.04 wt.%	Laminar and Turbulent	Heat transfer and pressure drop were same for all concentrations used
[12]	Counter flow double pipe	NDG/H_2O		Triton X-100	0.01–0.06 wt.%	Turbulent	15.87% enhancement of heat transfer coefficient was noticed at 15,000 Re number and 0.06 wt.%
[13]	Double pipe with and without helical coils	TiO_2/EG-H_2O	21 nm	Oleic acid and CTAB	0.0004–0.02 vol.%	Turbulent	Heat transfer was enhanced to 17.71% at 15,000 Re number and P/d ratio 2.5
[14]	Double pipe with louvered strips	Al_2O_3, CuO, SiO_2, ZnO/H_2O	20–50 nm		1–4 vol.%		SiO_2 showed maximum Nusselt number and skin fiction coefficient than other nanofluids

(continued)

Table 3.1 (continued)

References	Heat exchanger	Nanofluid	Nanoparticle size	Surfactant	Concentration	Flow	Outcomes
[15]	Double pipe	Al_2O_3/Transformer oil	7, 43 and 27–42 nm		0.5 vol.%	Laminar	Maximum increase in heat transfer coefficient noticed was 25% for rod shaped alumina nanofluid in comparison with base fluid
[16]	Counter and parallel flow double pipe	Al_2O_3/H_2O	30 nm		0–10 vol.%	Turbulent	30% highest thermal efficiency and 32.7% enhancement in Nusselt number was achieved for counter flow heat exchanger
[17]	Counter current double pipe	$Ag/EG\text{-}H_2O$	40–50 nm	Not used	0.1–1 vol.%	Laminar-Turbulent	Heat transfer coefficient was increased by 67% at 1% volume fraction
[18]	Double pipe	Al_2O_3/H_2O	–	–	0.01–0.05 vol.%		Nusselt number was enhanced by 57.70% at 0.03 volume fraction and 50–150 Re number
[19]	Counter flow double pipe	$Mn_{0.6}Zn_{0.4}Fe_2O_4/H_2O$	10–30 nm		5 vol.%	Turbulent for Coolant and Laminar for hot fluid	Pressure drop and heat transfer was improved by enhancing magnetic field, volume concentration and particle size

(continued)

Table 3.1 (continued)

References	Heat exchanger	Nanofluid	Nanoparticle size	Surfactant	Concentration	Flow	Outcomes
[20]	Counter current double pipe	MWCNT/H_2O	3–5 nm	CTAB	0.04–0.25 wt.%		35% increase in HTC was achieved at 0.04% mass fraction with 3 porous plates
[21]	Double pipe	MgO/EG	50 nm	Nonylphenol ethoxylate	0.1–0.3 wt.%	Laminar-Turbulent	55.4, 35 and 19% augmentation in thermal conductivity, pressure drop and friction factor was observed at 0.3% weight concentration of MgO/EG nanofluid
[22]	Double pipe	Al_2O_3–TiO_2/H_2O	30–40 nm	Polyvinyl Alcohol	0.025–0.1 vol.%		Stability observed for hybrid nanofluid was 32 h
[23]	Double pipe U-bend	Fe_3O_4/H_2O		CTAB	0.005–0.06 vol.%	Turbulent	0.472 NTU was attained at 28,901 Re number, 0.06% volume concentration and 1 pitch ratio
[24]	Counter flow double pipe with porous media	Al_2O_3/H_2O	38 nm		3 vol.%	Laminar	At Re = 2000 and Da = 0.1, PEC achieved was 4.6

(continued)

Table 3.1 (continued)

References	Heat exchanger	Nanofluid	Nanoparticle size	Surfactant	Concentration	Flow	Outcomes
[25]	Mini channel double pipe	Fe_3O_4-CNT/H_2O	0.1–0.9		0.1–0.9 Fe_3O_4, 0–1.35 CNT vol.%		Viscosity of Newtonian fluid was higher at central region while non-Newtonian fluid had higher viscosity near tube wall
[26]	Double pipe	AC-Solar Glycol		GA	0.2–0.6 vol.%		Maximum TPF achieved was 1.487 at 0.2% concentration and 0.04 kg/sec flow rate
[27]	Double pipe	MWCNT/Solar Glycol	20–30	GA	0.2–0.6 vol.%	Laminar	CHTC was enhanced to maximum of 115% at 0.6% concentration and 0.04 kg/s flow rate
[28]	Double pipe	ZnO, CuO, TiO_2, BeO/H_2O				Laminar-Turbulent	1.55, 1.53, 1.51, and 1.49 thermal performance factor was achieved for TiO_2, BeO, ZnO and CuO, respectively
[29]	Double pipe with U-bend	Fe_3O_4/H_2O		CTAB	0.005–0.06 vol.%		Nusselt number by 38.75% at 0.06% concentration, 10 twist ratio and 30,000 Re number respectively

(continued)

Table 3.1 (continued)

References	Heat exchanger	Nanofluid	Nanoparticle size	Surfactant	Concentration	Flow	Outcomes
[30]	Flexible double pipe	$MWCNT/H_2O$		CTAB	0.04–0.25 wt.%		100% increment in heat transfer coefficient was observed at minimum mass fraction 0.04% and maximum vibration level 9 m/s^2
[31]	Double pipe	$Al_2O_3–TiO_2/H_2O$	> 40 nm	Not used	0.1–1.5 vol.%	Turbulent	Twisted tape had more positive effect on heat transfer rate than nanofluid
[32]	Double pipe with double twisted tapes	TiO_2/H_2O			0.05 vol.%		Maximum TPF of TiO_2/H_2O nanofluid attained was 1.33, 1.49 and 1.65 at 2.5, 2 and 1.5 twist ratio, respectively
[33]	Counter current double pipe	$COOH-CNT/H_2O$	10–20 nm	Nonylphenol ethoxylate	0.1–0.3 wt.%	Laminar and Turbulent	Thermal conductivity and thermal performance were enhanced up to 56 and 44% at 0.3% weight concentration of CNT/H_2O nanofluid

(continued)

Table 3.1 (continued)

References	Heat exchanger	Nanofluid	Nanoparticle size	Surfactant	Concentration	Flow	Outcomes
[34]	Sinusoidal-double pipe	Fe_3O_4/water			4 vol.%		Nu number was increased to 25% by raising shape coefficient of sinusoidal HE from 0.2 to 0.3
[35]	Double pipe with U bend	Fe_3O_4/H_2O	36 nm	CTAB	0.005–0.06 vol.%	Turbulent	Friction factor was 1.267 times greater than water at Re = 28,954, 0.06% volume concentration and AR = 1
[36]	Double pipe with helical baffle	Cu, CuO, CNT/H_2O			1–5 vol.%		Nu number was directly and inversely related to Cu/water, CuO/water and CNT/water volume concentration
[37]	Double pipe	γ-Al_2O_3/H_2O		Not used	0.05 and 0.15 vol.%	Turbulent	23 and 25% enhancement in heat transfer coefficient and friction factor was observed at 0.15% volume concentration, respectively

(continued)

Table 3.1 (continued)

References	Heat exchanger	Nanofluid	Nanoparticle size	Surfactant	Concentration	Flow	Outcomes
[38]	Double pipe	AlN/EG	30 nm		1–4 vol.%	Laminar	Nu number and friction factor was enhanced by 35 and 12.5% using hybrid AlN/EG nanofluid
[39]	Counter current horizontal double pipe	Fe_3O_4/H_2O			4 vol.%	Laminar	Optimum magnetic field found was between 1.33×10^6 and 2.37×10^6
[40]	Double pipe	n-Nonadecane/H_2O	55–175 nm	NP-10	0–1.68 vol.%	Turbulent	Improvement ratio achieved was 1.05–1.09 at 1.68% volume fraction for n-nonadecane
[41]	Mini double pipe	Al_2O_3/H_2O	20 nm		0.1–0.3 vol.%	Turbulent	7.32% enhancement in heat transfer rate was observed at 60 g/s flow rate using Al_2O_3/H_2O nanofluid
[42]	Double pipe	MgO/H_2O	10–60 nm		0.1–1 vol.%	Turbulent	3122.08 W/m^2 K HTC was achieved at Re = 28,000, concentration = 1% and twist ratio = 2.6

(continued)

Table 3.1 (continued)

References	Heat exchanger	Nanofluid	Nanoparticle size	Surfactant	Concentration	Flow	Outcomes
[43]	Double pipe with swirl effect	Al_2O_3, ZnO, TiO_2, CuO	20–50 nm		0–3 vol.%	Turbulent	Average heat transfer coefficient was raised by 14% at 3% volume concentration of A_2O_3/H_2O nanofluid
[44]	Double pipe	Ag–MoS_2 and Fe_3O_4–SiO_2			1 vol.%	Turbulent	62.21% enhancement in heat transfer coefficient was noted for spherical tube using Ag–MoS_2 nanoparticle at 1% concentration
[45]	Double tube Helical HE	Ag, TiO_2, Au, Cu, Al_2O_3 and Fe			2–8 vol.%		Among studied nanofluids, Au showed better thermal performance compared to other nanofluids
[46]	Double pipe with twisted tape on outer tube	CuO, Al_2O_3, SiO_{2s}	29 nm		2–4 vol.%		6.4% enhancement in thermal performance was observed using CuO/H_2O nanofluid at 4 vol.%
[47]	Double pipe with turbulators	Ag-Graphene and MWCNT-Fe_2O_3/H_2O			0.1–0.7 vol.%	Laminar	Ag-Graphene nanofluid showed better thermal performance compared to MWCNT-Fe_2O_3

(continued)

Table 3.1 (continued)

References	Heat exchanger	Nanofluid	Nanoparticle size	Surfactant	Concentration	Flow	Outcomes
[48]	Double pipe with twists	CuO/H_2O			0.5–3 mass.%	Laminar	Maximum PEC observed was 1.236 at Re = 2000 and mass fraction of 3%
[49]	Double pipe with gear turbulators	Al_2O_3/H_2O			1–6 vol.%	Laminar-Turbulent	Nu number, effectiveness and number of transfer units were enhanced by 1.21, 1.19 and 1.20 times using Al_2O_3/H_2O nanofluid along with disc turbulators
[50]	Double pipe with fins	Al, Cu, CuO, Ti, TiO_2, Graphene, MWCNT and Al_2O_3	1–100 nm		1 vol.%	Transition-Turbulent	It was recommended to use MWCNT or graphene nanofluid for high heat transfer and optimized cost in a finned double pipe heat exchanger
[88]	Counter flow double pipe	CuO/H_2O			0.5 and 1 vol.%		Maximum effectiveness and heat transfer coefficient achieved was 54 and 55.5% at 2 LPM flow rate and 1 vol.% concentration of nanofluid

(continued)

Table 3.1 (continued)

References	Heat exchanger	Nanofluid	Nanoparticle size	Surfactant	Concentration	Flow	Outcomes
[51]	Circular tube	Al_2O_3/H_2O	20 nm	Not used	0.2–2.5 vol.%	Laminar	Heat transfer coefficient ratio was enhanced to 1.41 at 6000 Pe number and 2.5% concentration
[52]	Circular tube	CuO/water	30–50 nm		> 0.24 vol.%	Turbulent	25 and 20% increment in heat transfer coefficient and pressure drop was observed at 0.3% volume concentration compared to water, respectively
[54]	Circular tube	Al_2O_3/H_2O	38 nm		1–6 vol.%	Turbulent	Heat transfer was enhanced with Re number and particle concentration
[59]	Circular tube	TiO_2/H_2O	30 nm		0.05–0.25 vol.%	Turbulent	22 and 25% enhancement in heat transfer coefficient and pressure drop was noticed at Re = 5000 and 0.25% concentration

(continued)

Table 3.1 (continued)

References	Heat exchanger	Nanofluid	Nanoparticle size	Surfactant	Concentration	Flow	Outcomes
[62]	Circular tube	γ-Al_2O_3/H_2O	20 nm		0.03–0.135 vol.%	Turbulent	Maximum increment in heat transfer coefficient was 48% at 0.054% concentration and 10,000 Re number
[63]	Circular tube	Cu/H_2O			0.2–2 vol.%	Laminar	27.8% increment in heat transfer coefficient was achieved at 3% volume concentration
[64]	Circular tube with twisted tapes	Al_2O_3/H_2O	47 nm	SDBS	0.02–0.5 vol.%	Turbulent	Heat transfer coefficient was enhanced by 42.17% at 22,000 Re number and 0.5% concentration for circular tube with twisted tape inserts (twist ratio 5) in comparison with water
[65]	Circular tube	Al_2O_3/H_2O	100 nm		1–4 vol.%	Laminar	Heat transfer was enhanced by enhancing concentration and wall shear stress
[66]	Circular tube	Al_2O_3/H_2O	10–50 nm		1–4 vol.%	Laminar	Nu number was reduced by raising the Al_2O_3 nanoparticle size

(continued)

Table 3.1 (continued)

References	Heat exchanger	Nanofluid	Nanoparticle size	Surfactant	Concentration	Flow	Outcomes
[67]	Circular tube	Al_2O_3/H_2O	47 nm	SDBS	0.1 vol.%	Transition	23.69% enhancement in heat transfer was observed at 0.1 vol.% and Re = 9000 in comparison with water
[69]	Circular tube	γ-Al_2O_3, TiO_2, CuO/DI H_2O	25 γ-Al_2O_3, 10 TiO_2, 30–50 CuO nm	CMC	0.1–1.5 vol.%	Turbulent	By changing Peclet number from 210,000 to 360,000, HTC was increased to 71, 68 and 67% for CuO/CMC, γ-Al_2O_3/CMC and TiO_2/CMC, respectively
[70]	Circular tube	Cu/H_2O	25 nm	Not used	0.2–2.5 vol.%	Laminar	HTCR was enhanced to 1.45 at 2% volume concentration and Pe = 6700
[71]	Circular tube	CuO/Base Oil	50 nm	Not used	0.2–2 wt.%	Laminar	Maximum enhancement of 12.7% in HTC was obtained at 2% mass fraction compared to base fluid

(continued)

Table 3.1 (continued)

References	Heat exchanger	Nanofluid	Nanoparticle size	Surfactant	Concentration	Flow	Outcomes
[74]	Circular tube	Al_2O_3–Cu/H_2O	15 nm	SLS	0.1 vol.%	Laminar	16.97 and 6% average enhancement in friction factor was observed for Al_2O_3–Cu/H_2O and Al_2O_3/H_2O nanofluid at 0.1 vol. %
[75]	Horizontal tube	TiO_2/H_2O	20–40 nm		0–4 vol.%	Turbulent	Maximum and minimum total entropy generation was found for Eulerian approach and single phase with increasing Re number, respectively
[76]	Circular tube	Al_2O_3, A'C'/H_2O	20–50 nm	Not used	3 vol.%	Laminar and Turbulent	Amorphous Carbonic nanofluid had thermal conductivity similar to that of water at 3.5% volume concentration
[77]	Circular tube	Al_2O_3/H_2O	19 and 38 nm		0–6 vol.%	Turbulent	It was suggested that base fluid (H_2O) was a better choice for maximizing energy efficiency

(continued)

Table 3.1 (continued)

References	Heat exchanger	Nanofluid	Nanoparticle size	Surfactant	Concentration	Flow	Outcomes
[78]	Circular tube	TiO_2/H_2O	15 nm		0.07–0.21 vol.%	Turbulent	Thermal performance was 4.2% greater than water at 0.21% TiO_2 concentration
[79]	Circular tube	Al_2O_3/H_2O	50–100 nm		1–4 vol.%	Laminar	Re number had greater impact on HTC as compared to volume concentration
[80]	Circular tube	Al_2O_3/H_2O	40 nm	Not used	0.1–2 vol.%	Laminar	Maximum pressure drop gained was 5.7 times greater than pure water
[81]	Circular tube with twisted tape	Al_2O_3/H_2O			0–0.5 vol.%	Laminar	Friction factor for twisted tape inserts was 1.0652 and 1.0413 times greater than water and plain tube at Re = 2200 and 0.5% volume concentration, respectively
[82]	Circular pipe	Graphene/H_2O	1.4–2.3 nm thickness, 270 nm-1.5 μm lateral size	PVA	0.005–0.02 vol.%	Laminar	Maximum increment of 14.2% in HTC was noted at Re = 1850 and 0.02% concentration

(continued)

Table 3.1 (continued)

References	Heat exchanger	Nanofluid	Nanoparticle size	Surfactant	Concentration	Flow	Outcomes
[83]	Circular tube	γ-Al_2O_3, TiO_2, CuO/CMC	25 γ-Al_2O_3, 10 TiO_2, 30–50 CuO nm		0.1–1.5 vol.%	Laminar	Local heat transfer coefficient was reduced by 26.4% during CuO/CMC nanofluid flow from inlet of tube to the outlet
[84]	Circular tube	γ-Al_2O_3, TiO_2, CuO/CMC	25 γ-Al_2O_3, 10 TiO_2, 30–50 CuO nm		0.1–1.5 vol.%	Turbulent	88.4, 86.7 and 87.7% enhancement in HTC was observed at 0.5 vol.% and Pe = 350,000 for TiO_2/CMC, CuO/CMC and γ-Al_2O_3/CMC nanofluid
[85]	Circular tube	Al_2O_3/H_2O	10–100 nm		0–6 vol.%	Turbulent	It was necessary to use low concentration of nanoparticle to diminish total entropy generation was minimized at constant velocity
[86]	Helical coiled	CuO, Al_2O_3, ZnO/H_2O			1–4 vol.%	Turbulent	7.14 and 6.14% enhancement and reduction in heat transfer coefficient and entropy generation was observed, respectively

(continued)

Table 3.1 (continued)

References	Heat exchanger	Nanofluid	Nanoparticle size	Surfactant	Concentration	Flow	Outcomes
[87]	Helical coiled		25 nm		4 vol.%	Laminar	29 and 34% enhancement in performance index and effectiveness was achieved for counter flow helical coiled heat exchanger

References

1. IEA. (2021). *Global energy review 2021*. IEA, Paris. https://www.iea.org/reports/global-energy-review-2021
2. Yang, R., & Chiang, F. P. (2002). An experimental heat transfer study for periodically varying-curvature curved-pipe. *International journal of heat and mass transfer, 45*(15), 3199–3204.
3. Zhang, L., et al. (2012). Compound heat transfer enhancement for shell side of double-pipe heat exchanger by helical fins and vortex generators. *Heat and Mass Transfer, 48*(7), 1113–1124.
4. Akpinar, E. K., & Bicer, Y. (2005). Investigation of heat transfer and exergy loss in a concentric double pipe exchanger equipped with swirl generators. *International Journal of Thermal Sciences, 44*(6), 598–607.
5. Akpinar, E. K. (2006). Evaluation of heat transfer and exergy loss in a concentric double pipe exchanger equipped with helical wires. *Energy Conversion and Management, 47*(18–19), 3473–3486.
6. Choudhari, S. S., & Taji, S. (2013). Experimental studies on effect of coil wire insert on heat transfer enhancement and friction factor of double pipe heat exchanger. *International Journal of Computer Engineering Research, 3*, 3239.
7. Yildiz, C., Bíçer, Y., & Pehlívan, D. (1996). Influence of fluid rotation on the heat transfer and pressure drop in double-pipe heat exchangers. *Applied Energy, 54*(1), 49–56.
8. Zohir, A., Habib, M., & Nemitallah, M. (2015). Heat transfer characteristics in a double-pipe heat exchanger equipped with coiled circular wires. *Experimental Heat Transfer, 28*(6), 531–545.
9. Sheikholeslami, M., Gorji-Bandpy, M., & Ganji, D. D. (2016). Experimental study on turbulent flow and heat transfer in an air to water heat exchanger using perforated circular-ring. *Experimental Thermal and Fluid Science, 70*, 185–195.
10. Zamzamian, A., et al. (2011). Experimental investigation of forced convective heat transfer coefficient in nanofluids of Al2O3/EG and CuO/EG in a double pipe and plate heat exchangers under turbulent flow. *35*(3), 495–502.
11. Wu, Z., Wang, L., & Sunden, B. (2013). Pressure drop and convective heat transfer of water and nanofluids in a double-pipe helical heat exchanger. *60*(1–2), 266–274.
12. Goodarzi, M., et al. (2016). Investigation of heat transfer performance and friction factor of a counter-flow double-pipe heat exchanger using nitrogen-doped, graphene-based nanofluids. *76*, 16–23.
13. Reddy, M. C. S., & Rao, V. V. (2014). Experimental investigation of heat transfer coefficient and friction factor of ethylene glycol water based TiO_2 nanofluid in double pipe heat exchanger with and without helical coil inserts. *50*, 68–76.
14. Mohammed, H., et al. (2013). Heat transfer enhancement of nanofluids in a double pipe heat exchanger with louvered strip inserts. *40*, 36–46.
15. Chun, B.-H., Kang, H. U., & Kim, S. H. (2008). Effect of alumina nanoparticles in the fluid on heat transfer in double-pipe heat exchanger system. *25*(5), 966–971.
16. Bahmani, M. H., et al. (2018). Investigation of turbulent heat transfer and nanofluid flow in a double pipe heat exchanger. *29*(2), 273–282.
17. Sarafraz, M., Hormozi, F. J. E. T., & Science, F. (2015). Intensification of forced convection heat transfer using biological nanofluid in a double-pipe heat exchanger. *66*, 279–289.
18. Shirvan, K. M., et al. (2017). Numerical investigation of heat exchanger effectiveness in a double pipe heat exchanger filled with nanofluid: A sensitivity analysis by response surface methodology. *313*, 99–111.
19. Bahiraei, M., & Hangi, M. (2013). Investigating the efficacy of magnetic nanofluid as a coolant in double-pipe heat exchanger in the presence of magnetic field. *76*, 1125–1133.
20. Moradi, A., et al. (2019). An experimental study on MWCNT–water nanofluids flow and heat transfer in double-pipe heat exchanger using porous media. *Journal of Thermal Analysis and Calorimetry, 137*, 1–11.
21. Arya, H., et al. (2019). Heat transfer and pressure drop characteristics of MgO nanofluid in a double pipe heat exchanger. *55*(6), 1769–1781.

22. Mund, A., et al. (2019). Experimental and numerical study of heat transfer in double-pipe heat exchanger using Al_2O_3, and TiO_2 water nanofluid. In *Advances in Fluid and Thermal Engineering* (pp. 531–540). Springer.
23. Sundar, L. S., et al. (2019). Heat transfer and effectiveness experimentally-based analysis of wire coil with core-rod inserted in Fe_3O_4/water nanofluid flow in a double pipe U-bend heat exchanger. *International Journal of Heat and Mass Transfer, 134*, 405–419.
24. Siavashi, M., & Joibary, S. M. M. (2019). Numerical performance analysis of a counter-flow double-pipe heat exchanger with using nanofluid and both sides partly filled with porous media. *Journal of Thermal Analysis and Calorimetry, 135*(2), 1595–1610.
25. Shahsavar, A., et al. (2019). Impact of variable fluid properties on forced convection of Fe_3O_4/CNT/water hybrid nanofluid in a double-pipe mini-channel heat exchanger. *Journal of Thermal Analysis and Calorimetry, 135*, 1–13.
26. Poongavanam, G. K., et al. (2019). Heat transfer and pressure drop performance of solar glycol/activated carbon based nanofluids in shot peened double pipe heat exchanger. *140*, 580–591.
27. Poongavanam, G. K., et al. (2019). Experimental investigation on heat transfer and pressure drop of MWCNT-Solar glycol based nanofluids in shot peened double pipe heat exchanger. *345*, 815–824.
28. Gnanavel, C., Saravanan, R., & Chandrasekaran, M. (2019). Heat transfer enhancement through nano-fluids and twisted tape insert with rectangular cut on its rib in a double pipe heat exchanger.
29. Kumar, N. R., et al. (2018). Effect of twisted tape inserts on heat transfer, friction factor of Fe_3O_4 nanofluids flow in a double pipe U-bend heat exchanger. *Journal of Thermal Analysis and Calorimetry, 95*, 53–62.
30. Hosseinian, A., et al. (2018). Experimental investigation of surface vibration effects on increasing the stability and heat transfer coeffcient of MWCNTs-water nanofluid in a flexible double pipe heat exchanger. *90*, 275–285.
31. Maddah, H., et al. (2018). Factorial experimental design for the thermal performance of a double pipe heat exchanger using Al_2O_3-TiO_2 hybrid nanofluid. *Journal of Thermal Analysis and Calorimetry, 97*, 92–102.
32. Eiamsa-ard, S., Ketrain, R., & Chuwattanakul, V. (2018). TiO_2/water nanofluid heat transfer in heat exchanger equipped with double twisted-tape inserts. In *IOP Conference Series: Materials Science and Engineering*. IOP Publishing.
33. Sarafraz, M., et al. (2016). Thermal performance of a counter-current double pipe heat exchanger working with COOH-CNT/water nanofluids. *78*, 41–49.
34. Mousavi, S. V., et al. (2016). The Influence of magnetic field on heat transfer of magnetic nanofluid in a sinusoidal double pipe heat exchanger. *113*, 112–124.
35. Kumar, N. R., et al. (2017). Heat transfer, friction factor and effectiveness of Fe_3O_4 nanofluid flow in an inner tube of double pipe U-bend heat exchanger with and without longitudinal strip inserts. *85*, 331–343.
36. Saeedan, M., et al. (2016). CFD Investigation and neutral network modeling of heat transfer and pressure drop of nanofluids in double pipe helically baffled heat exchanger with a 3-D fined tube. *Applied Thermal Engineering, 100*, 721–729.
37. Raei, B., et al. (2017). Experimental study on the heat transfer and flow properties of γ-Al_2O3/water nanofluid in a double-tube heat exchanger. *Journal of Thermal Analysis and Calorimetry, 127*(3), 2561–2575.
38. Hussein, A. M. (2017). Thermal performance and thermal properties of hybrid nanofluid laminar flow in a double pipe heat exchanger. *Experimental Thermal and Fluid Science, 88*, 37–45.
39. Shakiba, A., & Vahedi, K. (2016). Numerical analysis of magnetic field effects on hydro-thermal behavior of a magnetic nanofluid in a double pipe heat exchanger. *Journal of Magnetism and Magnetic Materials, 402*, 131–142.
40. Doruk, S., et al. (2017). Heat transfer performance of water and Nanoencapsulated n-nonadecane based Nanofluids in a double pipe heat exchanger. *53*(12), 3399–3408.

41. Aghayari, R., et al. (2015). *Effect of nanoparticles on heat transfer in mini double-pipe heat exchangers in turbulent flow*. Springer.
42. Ghasemi, N., et al. (2019). Proposing a method for combining monitored multilayered perceptron (MLP) and self-organizing map (SOM) neural networks in prediction of heat transfer parameters in a double pipe heat exchanger with nanofluid. *Heat and Mass Transfer, 55*, 1–16.
43. Badawy, G. H., et al. (2022). Effect of nanofluids on the thermal performance of double pipe heat exchanger. *ERJ Engineering Research Journal, 45*(1), 13–25.
44. Asadi, A., et al. (2022). Numerical analysis of turbulence-inducing elements with various geometries and utilization of hybrid nanoparticles in a double pipe heat exchanger. *Alexandria Engineering Journal, 61*(5), 3633–3644.
45. Kerur, S. M., et al. (2022). Numerical investigation and performance comparison of double-tube helical heat exchanger incorporating four-turn and six-turn models using several nanofluids. In *Recent advances in mechanical infrastructure* (pp. 31–43). Springer.
46. Noorbakhsh, M., et al. (2022). Thermal analysis of nanofluids flow in a double pipe heat exchanger with twisted tapes insert in both sides. *Journal of Thermal Analysis and Calorimetry, 147*(5), 3965–3976.
47. Hashemi Karouei, S. H., et al. (2021). Laminar heat transfer and fluid flow of two various hybrid nanofluids in a helical double-pipe heat exchanger equipped with an innovative curved conical turbulator. *Journal of Thermal Analysis and Calorimetry, 143*(2), 1455–1466.
48. Shahsavar, A., et al. (2021). Numerical study of the possibility of improving the hydrothermal performance of an elliptical double-pipe heat exchanger through the simultaneous use of twisted tubes and non-Newtonian nanofluid. *Journal of Thermal Analysis and Calorimetry, 143*(3), 2825–2840.
49. Bashtani, I., Esfahani, J. A., & Kim, K. C. (2021). Effects of water-aluminum oxide nanofluid on double pipe heat exchanger with gear disc turbulators: A numerical investigation. *Journal of the Taiwan Institute of Chemical Engineers, 124*, 63–74.
50. Dalkılıç, A. S., et al. (2021). Optimization of the finned double-pipe heat exchanger using nanofluids as working fluids. *Journal of Thermal Analysis and Calorimetry, 143*(2), 859–878.
51. Heris, S. Z., et al. (2007). Experimental investigation of convective heat transfer of Al_2O_3/water nanofluid in circular tube. *International Journal of Heat and Fluid Flow, 28*(2), 203–210.
52. Fotukian, S., & Esfahany, M. N. (2010). Experimental study of turbulent convective heat transfer and pressure drop of dilute CuO/water nanofluid inside a circular tube. *International Communications in Heat and Mass Transfer, 37*(2), 214–219.
53. El Bécaye Maïga, S., et al. (2006). Heat transfer enhancement in turbulent tube flow using Al_2O_3 nanoparticle suspension. *International Journal of Numerical Methods for Heat & Fluid Flow, 16*(3), 275–292.
54. Bianco, V., Manca, O., & Nardini, S. (2011). Numerical investigation on nanofluids turbulent convection heat transfer inside a circular tube. *International Journal of Thermal Sciences, 50*(3), 341–349.
55. Das, S. K., et al. (2007). *Nanofluids: Science and technology*. Wiley.
56. Behzadmehr, A., et al. (2007). Prediction of turbulent forced convection of a nanofluid in a tube with uniform heat flux using a two phase approach. *28*(2), 211–219.
57. Mirmasoumi, S., & Behzadmehr, A. (2008). Numerical study of laminar mixed convection of a nanofluid in a horizontal tube using two-phase mixture model. *28*(7), 717–727.
58. Akbarinia, A., & Laur, R. (2009). Flow, Investigating the diameter of solid particles effects on a laminar nanofluid flow in a curved tube using a two phase approach. *International Journal of Heat and Fluid Flow, 30*(4), 706–714.
59. Sajadi, A., & Kazemi, M. (2011). Investigation of turbulent convective heat transfer and pressure drop of TiO_2/water nanofluid in circular tube. *International Communications in Heat and Mass Transfer, 38*(10), 1474–1478.
60. Pak, B. C., & Cho, Y. I. (1998). Hydrodynamic and heat transfer study of dispersed fluids with submicron metallic oxide particles. *Experimental Heat Transfer, 11*(2), 151–170.
61. Xuan, Y., & Li, Q. (2000). Heat transfer enhancement of nanofluids. *International Journal of Heat and Fluid Flow, 21*(1), 58–64.

62. Fotukian, S., & Esfahany, M. N. (2010). Experimental investigation of turbulent convective heat transfer of dilute γ-Al_2O_3/water nanofluid inside a circular tube. *International Journal of Heat and Fluid Flow, 31*(4), 606–612.
63. Fard, M. H., et al. (2010). Numerical study of convective heat transfer of nanofluids in a circular tube two-phase model versus single-phase model. *37*(1), 91–97.
64. Sundar, L. S., & Sharma, K. (2010). Turbulent heat transfer and friction factor of Al_2O_3 nanofluid in circular tube with twisted tape inserts. 2010. *International Journal of Heat and Mass Transfer, 53*(7–8), 1409–1416.
65. Bianco, V., et al. (2009). Numerical investigation of nanofluids forced convection in circular tubes. *29*(17–18), 3632–3642.
66. Heris, S. Z., Esfahany, M. N., & Etemad, G. (2007). Numerical investigation of nanofluid laminar convective heat transfer through a circular tube. *Numerical Heat Transfer, Part A: Applications, 52*(11), 1043–1058.
67. Sharma, K., et al. (2009). Estimation of heat transfer coefficient and friction factor in the transition flow with low volume concentration of Al_2O_3 nanofluid flowing in a circular tube and with twisted tape insert. *International Communications in Heat and Mass Transfer, 36*(5), 503–507.
68. Gnielinski, V. (1976). New equations for heat and mass transfer in turbulent pipe and channel flow. *16*(2), 359–368.
69. Hojjat, M., et al. (2011). Convective heat transfer of non-Newtonian nanofluids through a uniformly heated circular tube. *50*(4), 525–531.
70. Heris, S. Z., Etemad, S. G., & Esfahany, M. N. (2009). Convective heat transfer of a Cu/water nanofluid flowing through a circular tube. *Experimental Heat Transfer, 22*(4), 217–227.
71. Saeedinia, M., et al. (2012). Thermal and rheological characteristics of CuO-base oil nanofluid flow inside a circular tube. *International Communications in Heat and Mass Transfer, 39*(1), 152–159.
72. Hamilton, R. L., & Crosser, O. (1962). Thermal conductivity of heterogeneous two-component systems. *Industrial & Engineering Chemistry Fundamentals, 1*(3), 187–191.
73. Yu, W., & Choi, S. (2003). The role of interfacial layers in the enhanced thermal conductivity of nanofluids: A renovated Maxwell model. *Journal of Nanoparticle Research, 5*(1–2), 167–171.
74. Suresh, S., et al. (2012). Effect of Al_2O_3–Cu/water hybrid nanofluid in heat transfer. *Experimental Thermal and Fluid Science, 38*, 54–60.
75. Rashidi, M. M., et al. (2017). Entropy generation in a circular tube heat exchanger using nanofluids: Effects of different modeling approaches. *38*(9), 853–866.
76. Kim, D., et al. (2009). Convective heat transfer characteristics of nanofluids under laminar and turbulent flow conditions. *9*(2), e119–e123.
77. Bianco, V., Manca, O., & Nardini, S. J. E. (2014). Performance analysis of turbulent convection heat transfer of Al_2O_3 water-nanofluid in circular tubes at constant wall temperature. *77*, 403–413.
78. Eiamsa-Ard, S., et al. (2015). Heat transfer enhancement of TiO_2/water nanofluid in a heat exchanger tube equipped with overlapped dual twisted-tapes. *18*(3), 336–350.
79. Tahir, S., & Mital, M. (2012). Numerical investigation of laminar nanofluid developing flow and heat transfer in a circular channel. *39*, 8–14.
80. Heyhat, M., et al. (2013). Experimental investigation of laminar convective heat transfer and pressure drop of water-based Al_2O_3 nanofluids in fully developed flow regime. *44*, 483–489.
81. Sundar, L. S., & Sharma, K. (2011). Laminar convective heat transfer and friction factor of Al_2O_3 nanofluid in circular tube fitted with twisted tape inserts. *International Journal of Automotive and Mechanical Engineering, 3*(2), 65–278.
82. Akhavan-Zanjani, H., et al. (2016). Experimental investigation of laminar forced convective heat transfer of Graphene–water nanofluid inside a circular tube. *100*, 316–323.
83. Hojjat, M., Etemad, S. G., & Bagheri, R. (2010). Laminar heat transfer of non-Newtonian nanofluids in a circular tube. *Korean Journal of Chemical Engineering, 27*(5), 1391–1396.
84. Hojjat, M., et al. (2011). Turbulent forced convection heat transfer of non-Newtonian nanofluids. *35*(7), 1351–1356.

85. Bianco, V., et al. (2014). Entropy generation analysis of turbulent convection flow of Al_2O_3–water nanofluid in a circular tube subjected to constant wall heat flux. *77*, 306–314.
86. Khairul, M. A., et al. (2013). Heat transfer and thermodynamic analyses of a helically coiled heat exchanger using different types of nanofluids. *67*, 398–403.
87. Mohammed, H. A., & Narrein, K. (2012). Thermal and hydraulic characteristics of nanofluid flow in a helically coiled tube heat exchanger. *International Communications in Heat and Mass Transfer, 39*(9), 1375–1383.
88. Hussein, A. M., & Danook, S. H. (2022). Enhancement of double-pipe heat exchanger effectiveness by using water-CuO. *NTU Journal of Engineering and Technology, 1*(2), 18–22.

Chapter 4
Microchannel Heat Exchanger

Abstract Microchannel heat exchangers are the future of various industries such as robotics, electronic devices, solar energy systems and aerospace. A brief introduction about use of nanofluid in microchannel heat exchanger has been discussed. Effect of Reynold number, channel geometry, number of channels and volume fraction on thermal performance of MCHE has been reviewed.

4.1 Introduction

In last few decades, microchannel heat exchanger (MCHE) has gained attention due to their compact size. Microchannel heat exchangers are generally 1 mm to 1 μm in length. These heat exchangers are widely used in aerospace, robotics, microelectronics and automotive industries [1]. Geometry of channel and its size significantly affect the heat transfer performance of MCHE.

Zhou et al. [2] evaluated thermal performance of MCHE experimentally and numerically and found that Nusselt number was raised by decreasing microchannel height and increasing microchannel width. Hydrothermal properties of microchannel was raised by introducing fan shaped cavities with decrement in pressure drop [3]. Pan et al. [4] used trapezoidal cavities in microchannel heat exchanger and compared its performance with conventional MCHE. It was found that thermal efficacy was raised by enhancing mass flow rate up to an optimum point which was found to be higher than conventional MCHE. Manifold MCHE was proposed by Zhang et al. [5] for aerospace applications. Inconel 718 was used to fabricate MCHE and N_2 was used at 600 °C as hot fluid while air was used as cold fluid at 38 °C. Result showed 25% greater heat transfer density compared to plate fin heat exchanger. Kwon et al. [6] studied two-phase cooling using air and refrigerant R245fa at a temperature of 80 °C in microchannel heat exchanger. Power density was found to be 100 W/cm^3 greater than other industrial heat exchangers. Sun et al. [7] found higher thermal performance of microchannel heat exchanger with 2 times higher heat transfer coefficient compared to fin tube heat exchanger. While, at the same time pressure drop was around 125% greater at same conditions. Hasan et al. [8] compared the performance of different shapes of microchannels (square, circular,

H. M. Ali et al., *Nanofluids for Heat Exchangers*,
https://doi.org/10.1007/978-981-19-3227-4_4

rectangular and trapezoidal) and number of channels. Result showed that heat transfer and pressure drop was increased by raising the number of channels. Among studied shapes, circular microchannel was found best in terms of performance index followed by square. Brandner et al. [9] enhanced performance of microchannel heat exchanger by reducing the hydraulic diameter of microchannels.

Besides optimizing the performance of MCHE through geometries, an alternative approach is to use thermal fluids with high heat transfer properties. Nanofluids can be used in MCHE to optimize its thermal performance due to their better thermal properties compared to conventional fluids.

4.2 Performance of Nanofluid in Microchannel Heat Exchanger

Particle size of nanofluid is crucial in enhancing performance of MCHE. Mohammadian et al. [10] used three-dimensional conjugate heat transfer model to analyze the particle size and concentration of nanofluid on the performance of counter flow heat exchanger. Effectiveness, and pumping power was raised by increasing the concentration and reducing the particle size. Also, frictional entropy was raised by enhancing Re number, volume concentration and reducing particle size. Seyf and Feizbakhshi [11] found that Nusselt number was increased by decreasing the particle size of Al_2O_3/H_2O nanofluid while the trend was opposite for CuO/H_2O nanofluid in a micro pin heat sink (Fig. 4.1).

Seyf and Feizbakhshi [11] also found that CuO/H_2O nanofluid had higher Nusselt number compared to Al_2O_3/H_2O nanofluid as shown in Fig. 4.2 because of higher thermal properties of CuO nanoparticles.

Das et al. [12] found that thermal conductivity was raised in MCHE by using smaller sized CuO/H_2O nanofluid. Nguyen et al. [13] used turbulent condition in an electronic cooling system using Al_2O_3/H_2O nanofluid. It was found that higher heat transfer was observed for particle size of 36 nm compared to 47 nm particle size of Al_2O_3. In another study, Anoop et al. [14] found that Al_2O_3/H_2O nanofluid had higher

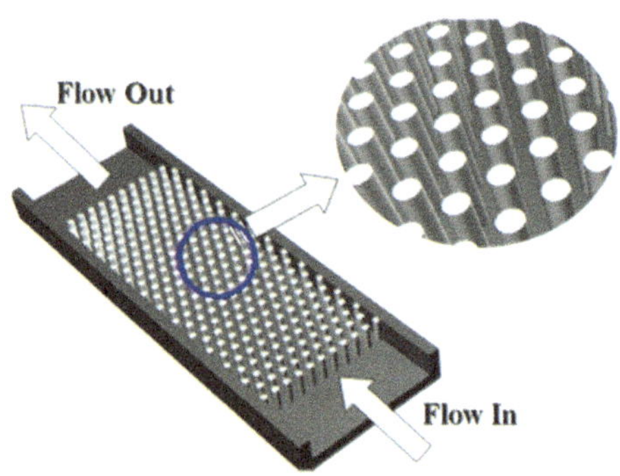

Fig. 4.1 Schematic of micro pin heat sink [11]

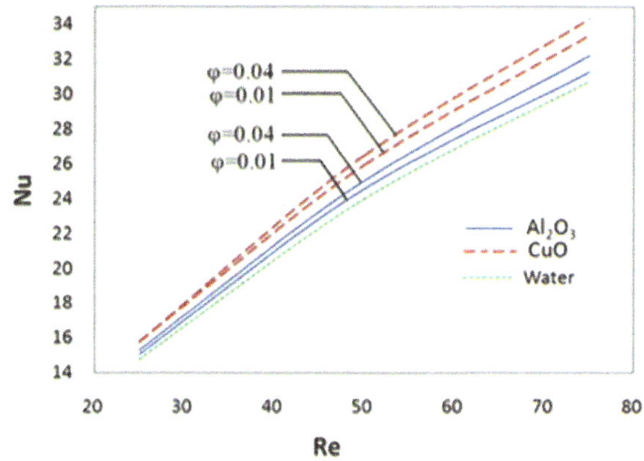

Fig. 4.2 Effect of nanofluid concentration and Reynold number on Nusselt number [11]

heat transfer coefficient at small particle size of 45 nm compared to 150 nm. While TiO_2/H_2O nanofluid particle size was less sensitive to heat transfer performance at constant Reynold number and volume concentration.

Mazaheri et al. [15] carried out two-phase analysis using nanofluid as hot fluid and water as cold fluid in 4-layer MCHE. Figure of 4-layer MCHE is shown in Fig. 4.3.

Performance evaluation criteria was increased by increasing the volume fraction of nanofluid while, it was reduced by increasing number of channels and mass flow rate. Overall heat transfer coefficient was increased with number of microchannels due to thermal boundary layer disruption (Fig. 4.4).

Li et al. [16] enhanced heat transfer coefficient of microchannel heat sink (MCHS) using carbon/acetone nanofluid. HTC was enhanced by 73% using nanofluid with

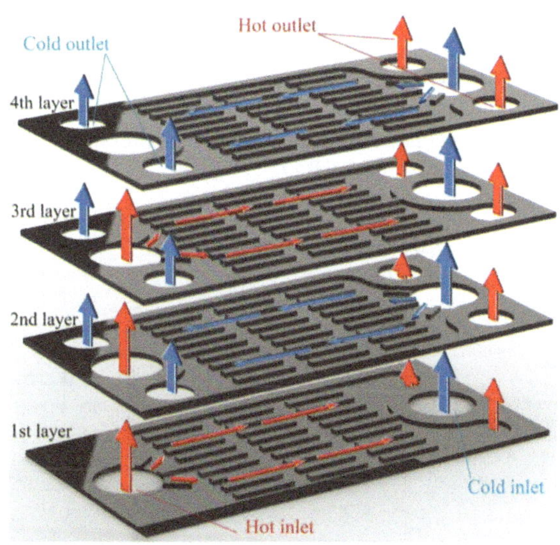

Fig. 4.3 4 layer microchannel heat exchanger [15]

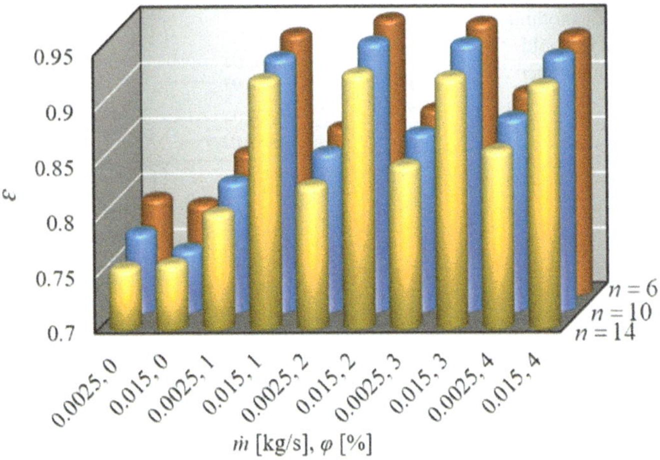

Fig. 4.4 Effect of mass flow rate, number of channels and volume fraction on effectiveness [15]

increment in friction factor and pressure drop. Thermal hydraulic performance was also enhanced by 69% due to thermophoresis and Brownian motion phenomenon in nanofluids and decrement in thermal boundary layer with in microchannel helped in enhancing heat transfer in MCHS (Fig. 4.5).

Comparison was made between Cu/EG and alumina/EG nanofluids in a microchannel heat exchanger. It was found that Cu/EG had better thermal hydraulic performance compared to Al_2O_3/EG nanofluid due to better thermal properties [17]. Sivakumar et al. [18] analyzed performance of serpentine shaped microchannel heat exchanger using CuO/H_2O and Al_2O_3/H_2O nanofluid under laminar flow. HTC was raised in nanofluid compared to base fluid by increasing volume fraction. Also, comparison between CuO and Al_2O_3 showed that CuO had higher thermal efficiency than Al_2O_3/H_2O nanofluid in MCHE. Ahlatli et al. [19] enhanced thermal hydraulic performance in solar microchannel collector using CNT/H_2O nanofluid at laminar condition at the expense of pressure drop. Omri et al. [20] numerically

Fig. 4.5 Effect of volume fraction and Reynold number of pressure drop [16]

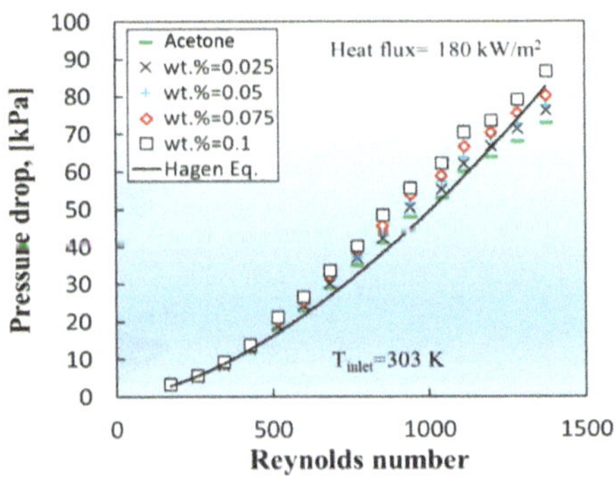

analyzed performance of microchannel HE equipped with triangular fins using finite element method and CNT/H_2O nanofluid. Nanofluid helped in enhancing thermal performance factor while height of fins in MCHE had inverse effect on thermal performance of MCHE. Figure 4.6 showed the schematic of triangular fin MCHE.

Mazaheri et al. [21] numerically evaluated the second law performance of MCHE using nanofluid. It was found that entropy generation was enhanced by raising the mass flow rate while it was reduced by increasing the volume fraction of nanofluid. It

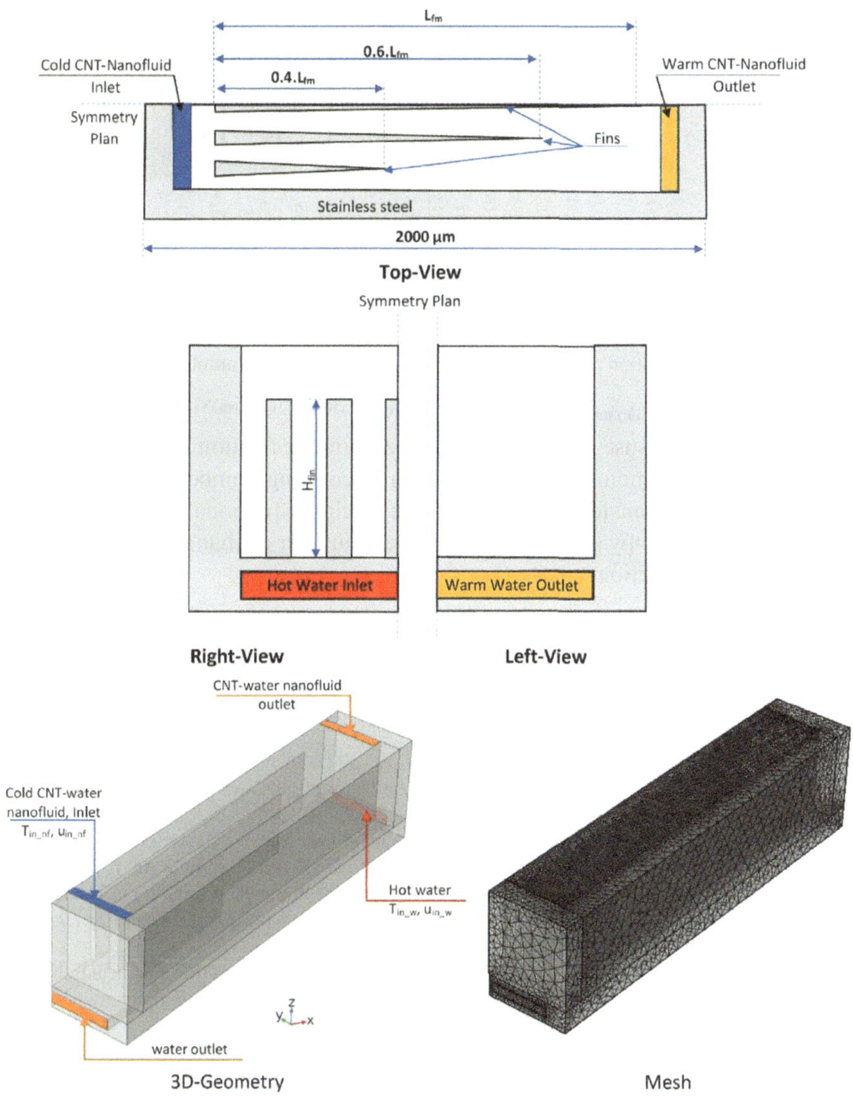

Fig. 4.6 Schematics of studied triangular fin MCHE [20]

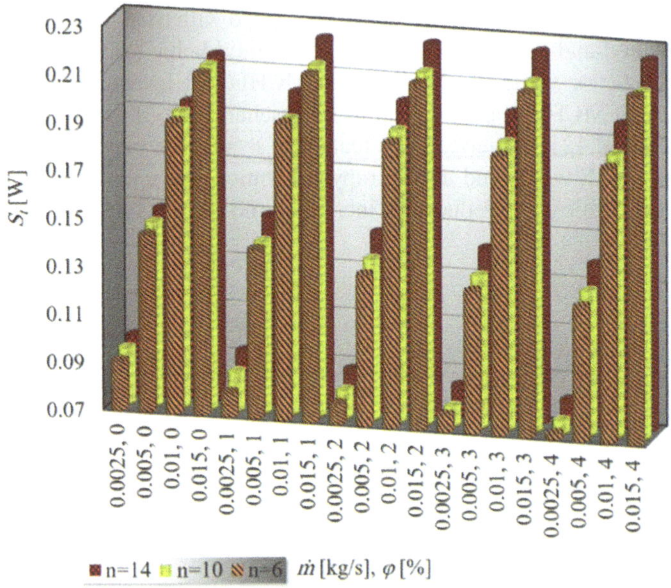

Fig. 4.7 Entropy generation at various flowrates, volume fraction and number of channels [21]

was recommended to use nanofluid with highest volume fraction and minimum flow rate to achieve minimum entropy generation. Increasing number of channels also caused higher heat transfer irreversibility due to higher frictional losses. Figure 4.7 showed the total entropy generation for different number of channels at various flow rates and volume fraction.

References

1. Bahrami, M., Yovanovich, M. M., & Culham, J. R. (2007). A novel solution for pressure drop in singly connected microchannels of arbitrary cross-section. *International Journal of Heat and Mass Transfer, 50*(13–14), 2492–2502.
2. Zhou, F., et al. (2018). Investigation of fluid flow and heat transfer characteristics of parallel flow double-layer microchannel heat exchanger. *Applied Thermal Engineering, 137*, 616–631.
3. Pan, M., et al. (2019). Experimental investigation of the heat transfer performance of microchannel heat exchangers with fan-shaped cavities. *International Journal of Heat and Mass Transfer, 134*, 1199–1208.
4. Pan, M., Zhong, Y., & Xu, Y. (2018). Numerical investigation of fluid flow and heat transfer in a plate microchannel heat exchanger with isosceles trapezoid-shaped reentrant cavities in the sidewall. *Chemical Engineering and Processing-Process Intensification, 131*, 178–189.
5. Zhang, X., et al. (2018). An additively manufactured metallic manifold-microchannel heat exchanger for high temperature applications. *Applied Thermal Engineering, 143*, 899–908.
6. Kwon, B., et al. (2019). High power density two-phase cooling in microchannel heat exchangers. *Applied Thermal Engineering, 148*, 1271–1277.

7. Sun, X., et al. (2019). Heat and mass transfer comparisons of desiccant coated microchannel and fin-and-tube heat exchangers. *Applied Thermal Engineering, 150*, 1159–1167.
8. Hasan, M. I., et al. (2009). Influence of channel geometry on the performance of a counter flow microchannel heat exchanger. *International Journal of Thermal Sciences, 48*(8), 1607–1618.
9. Brandner, J., et al. (2006). Concepts and realization of microstructure heat exchangers for enhanced heat transfer. *Experimental Thermal and Fluid Science, 30*(8), 801–809.
10. Mohammadian, S. K., Reza Seyf, H., & Zhang, Y. (2014). Performance augmentation and optimization of aluminum oxide-water nanofluid flow in a two-fluid microchannel heat exchanger. *Journal of Heat Transfer, 136*(2).
11. Seyf, H. R., & Feizbakhshi, M. (2012). Computational analysis of nanofluid effects on convective heat transfer enhancement of micro-pin-fin heat sinks. *International Journal of Thermal Sciences, 58*, 168–179.
12. Das, S. K., et al. (2003). Temperature dependence of thermal conductivity enhancement for nanofluids. *Journal of Heat Transfer, 125*(4), 567–574.
13. Nguyen, C. T., et al. (2007). Heat transfer enhancement using Al_2O_3–water nanofluid for an electronic liquid cooling system. *Applied Thermal Engineering, 27*(8–9), 1501–1506.
14. Anoop, K., Sundararajan, T., & Das, S. K. (2009). Effect of particle size on the convective heat transfer in nanofluid in the developing region. *International Journal of Heat and Mass Transfer, 52*(9–10), 2189–2195.
15. Mazaheri, N., Bahiraei, M., & Razi, S. (2021). Two-phase analysis of nanofluid flow within an innovative four-layer microchannel heat exchanger: Focusing on energy efficiency principle. *Powder Technology, 383*, 484–497.
16. Li, Z., et al. (2020). Heat transfer evaluation of a micro heat exchanger cooling with spherical carbon-acetone nanofluid. *International Journal of Heat and Mass Transfer, 149*, 119124.
17. Sohel, M., et al. (2013). Analysis of entropy generation using nanofluid flow through the circular microchannel and minichannel heat sink. *International Communications in Heat and Mass Transfer, 46*, 85–91.
18. Sivakumar, A., Alagumurthi, N., & Senthilvelan, T. (2016). Experimental investigation of forced convective heat transfer performance in nanofluids of Al_2O_3/water and CuO/water in a serpentine shaped micro channel heat sink. *Heat and Mass Transfer, 52*(7), 1265–1274.
19. Ahlatli, S., et al. (2016). Thermal performance of carbon nanotube nanofluids in solar microchannel collectors: An experimental study. *International Journal of Technology (IJTech), 2*, 78–85.
20. Omri, M., et al. (2022). A new microchannel heat exchanger configuration using CNT-nanofluid and allowing uniform temperature on the active wall. *Case Studies in Thermal Engineering, 32*, 101866.
21. Mazaheri, N., Bahiraei, M., & Razi, S. (2022). Second law performance of a novel four-layer microchannel heat exchanger operating with nanofluid through a two-phase simulation. *Powder Technology, 396*, 673–688.

Chapter 5
Preparation and Stability of Nanofluid

Abstract Nanofluid are the superior thermal fluids compared to conventional fluids. Only issue for using nanofluid is their preparation and stability for long term. In this chapter, preparation techniques for different types of nanofluids have been discussed and how stability of these nanofluids can be enhanced using different techniques such as use of surfactant.

5.1 Mono Nanofluids

Nanofluids can be considered as best replacement for conventional thermal fluids due to their superior thermal properties. Before using nanofluids in any thermal system, certain problems such as preparation of nanofluids and stability needs to be resolved. There are certain ways to resolve each issue as shown in Fig. 5.1.

Nanofluid can be prepared in two ways:

i. One Step Method
ii. Two Step Method

One Step Method

One step method involves simultaneous synthesis of nanoparticles and preparation of nanofluid in one step. One step method is quite expensive but it helps in avoiding transportation, storage and drying of nanoparticles. One step method involves three categories for preparation of nanofluid given in Figs. 5.2 and 5.3.

Vapor Deposition Method

Vapor deposition is the most used single step method developed by Choi and Eastman [3]. It involves a rotating disc which creates centrifugal force, this centrifugal force helps in forming a base fluid thin layer. Heated material is then evaporated into container containing inert gas at low pressure. Raw material will be condensed when encounters thin film of base fluid and then settled in it, nanofluid will be separated at the end (Fig. 5.4).

H. M. Ali et al., *Nanofluids for Heat Exchangers*,
https://doi.org/10.1007/978-981-19-3227-4_5

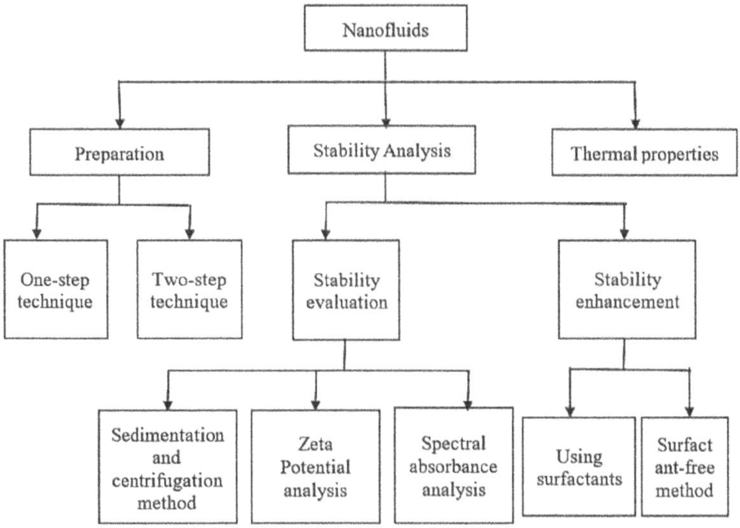

Fig. 5.1 Nanofluid representation in terms of preparation method, thermal properties and stability [1]

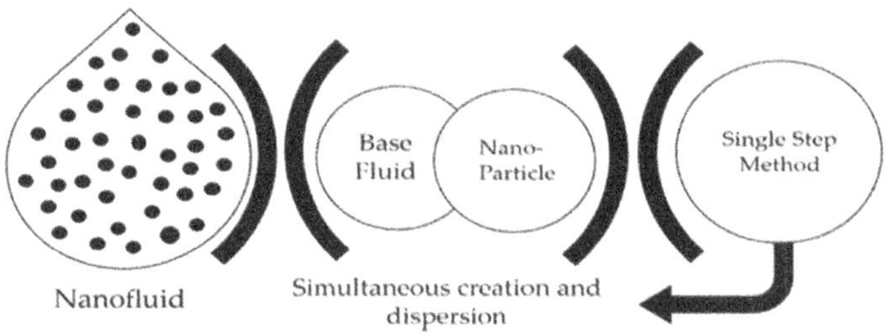

Fig. 5.2 One step method [2]

Submerged Arc Method

In a vacuum chamber, desired metal rod is submerged into the desired based fluid. Parameters such as base fluid temperature, pulse duration, voltage, current breakdown, off time duration and pressure inside the vacuum chamber is controlled. High pulsed arc of temperature 5000 to 20,000 K will be generated with the help of electric arc generator. High temperature arc evaporates the submerged metal rod along with base fluid. Metal particles will then be cooled down and condensed inside the vacuum chamber with the assistance of isolated liquid (Fig. 5.5).

Two Step Method

This is an extensively used method for the preparation of nanofluid. In this method, nanoparticles, nanosheets, nanotubes, and nanorods are synthesized separately using

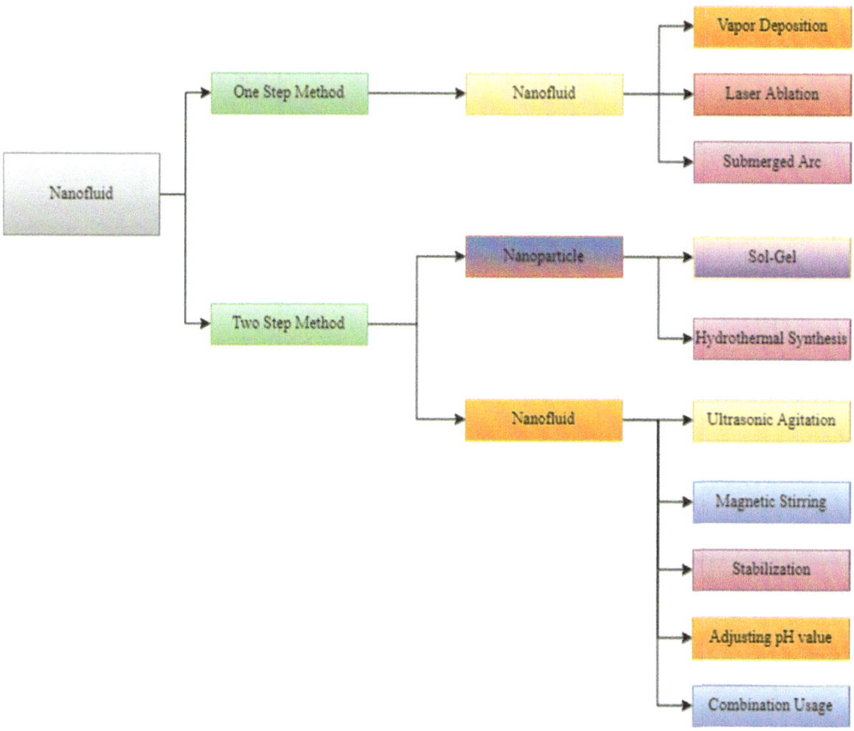

Fig. 5.3 Classification of preparation methods for nanofluids [2]

Fig. 5.4 Schematic diagram of vapor deposition method [2]

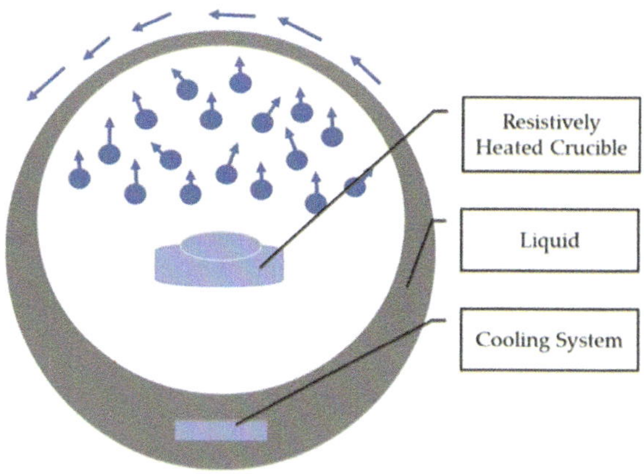

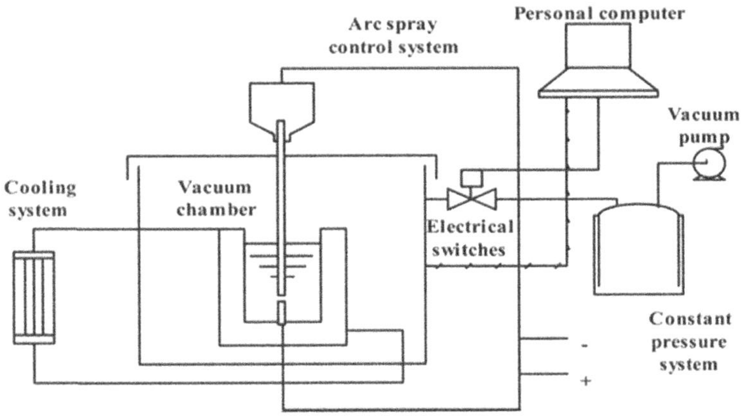

Fig. 5.5 Schematic of submerged arc method [2]

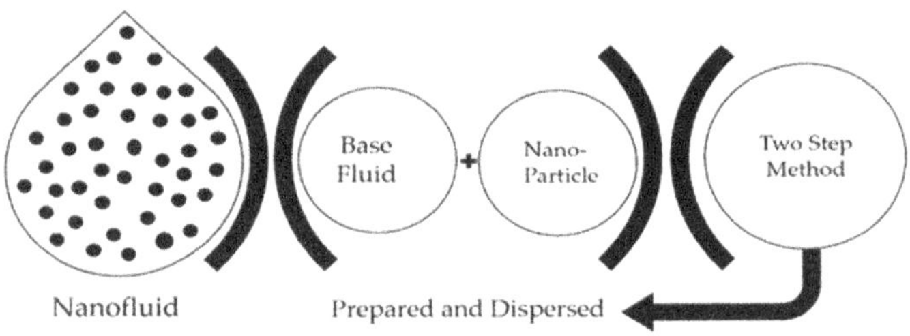

Fig. 5.6 Two step method [2]

hydrothermal synthesis, sol gel method or microemulsion methods. Synthesized nanoparticle will then be dispersed into base fluid using various techniques mentioned in Fig. 5.6.

Enhanced heat transfer properties are achieved by using a stable nanofluid. Stability of nanofluid plays an essential role in heat transfer. Unstable nanofluids reduced performance of system. It is necessary to utilize different techniques to prepare a stable nanofluid. Sun et al. [4] explained general preparation of nanofluid in Fig. 5.7.

Preparation of Graphene/H_2O

Graphene nanofluids have special characteristics that are essential for better thermal applications which includes the high thermal conductivity of graphene nanoparticles, reduced corrosion, erosion and clogging in the systems, better lubrication due to special structure. Graphene nanofluids can be prepared according to the graphene type such as graphene oxide, reduced graphene oxide and graphene. Chemical structure of all three is given in Fig. 5.8.

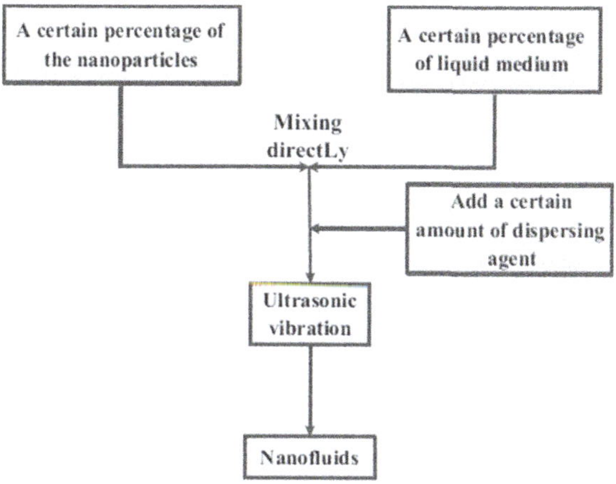

Fig. 5.7 Preparation of nanofluid by two step process [4]

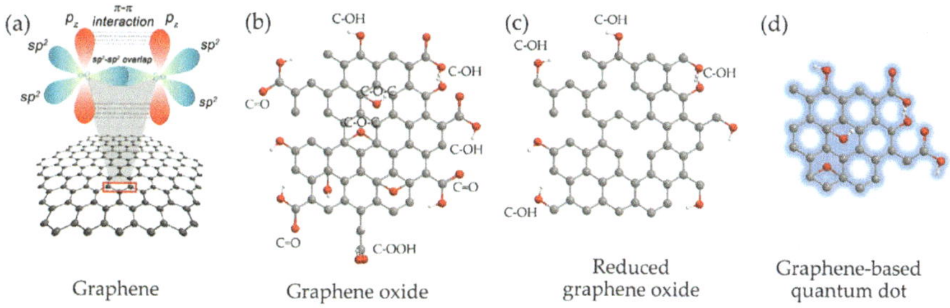

Fig. 5.8 Chemical structure of **a** Graphene. **b** Graphene oxide. **c** Reduced graphene oxide. **d** Graphene based quantum dot [5]

Synthesis of graphene nanoparticles is a challenging issue. Hummer's method is one of the way to synthesize graphene oxide nanoparticles from graphene. Esfahani et al. [6] used hummer's method to synthesize graphene oxide (synthesis given in Fig. 5.9). Synthesized nanoparticles were used to prepare graphene oxide–water nanofluid by stirring solution for 5 h and sonicated for 1 h at 42 kHz and 130 W.

Ghozatloo et al. [7] composed graphene nanosheets by employing catalytic decomposition method (CVD) over Cu foil. Graphene hydrophilic structure in H_2O was obtained by oxidizing it with alkaline media. Suspended alkaline graphene in water was sonicated for 15 min at different concentrations. Two step method was used for preparing graphene/H_2O nanofluid. Heat capacity of nanofluid was reduced by 8.11% at 0.1% weight fraction which was minor and heat capacity was like that of water.

In another study, graphene water nanofluid was prepared in a two step method. Gum Arabic was used as a surfactant to ensure enhanced stability of nanofluid. GA was mixed with deionized water prior to the graphene nanoparticles mixing.

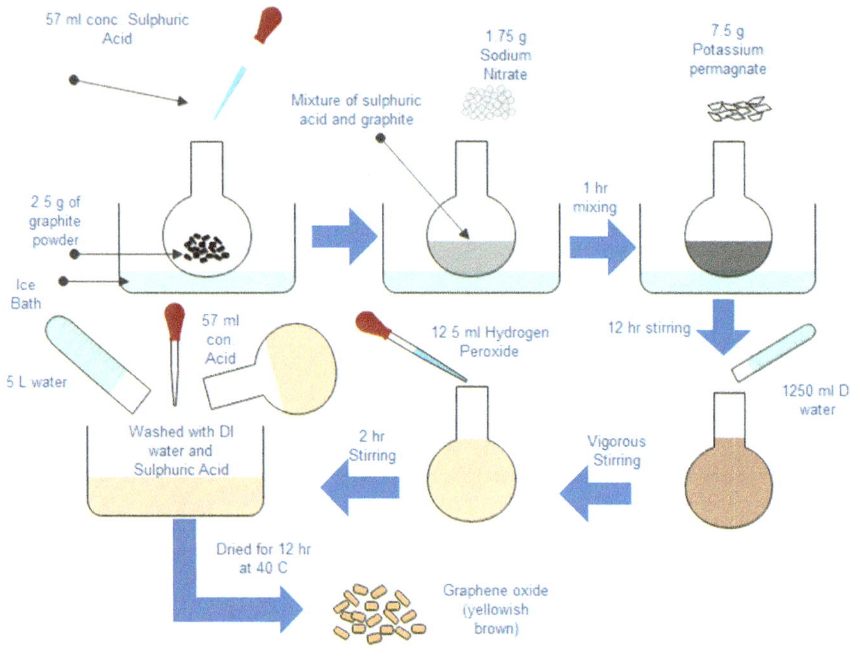

Fig. 5.9 Steps and procedure involved in synthesis of graphene oxide

Magnetic agitation was used for 15 min for more stable dispersion. Concentration and size of nanoparticles used was 0.01–0.2 wt% and 50–140 nm particle size [8].

Wang et al. [9] used one step process to prepare graphene/H_2O-EG nanofluid with Triton X-100 as surfactant. Graphene nanoparticles, Triton X-100 and EG-H_2O (50:50) was stirred for half an hour and then ultrasonication was done for 2 h. Graphene nanoparticles can be prepared by using different methods such as chemical vapor deposition, micromechanical cleavage and solvothermal synthesis. Preferable method among these was chemical method because this method had potential of bulk production [10].

Baby and Ramaprabhu [11] prepared graphene/H_2O-EG nanofluid without using any surfactant. Graphene nanoparticles were synthesized from graphene oxide in argon atmosphere at 1050 °C for 30 s. Prepared functionalized graphene nanoparticles were dispersed in H_2O and EG at specific concentration with the help of ultrasonicator for 30 to 45 min at a frequency of 45 kHz and maintaining the pH around 6–7. Figure 5.10 shows the effect of volume concentration and temperature on thermal conductivity of Graphene/H_2O and Graphene/EG nanofluid. Thermal conductivity was raised by increasing the volume fraction due to increased number of graphene nanoparticles in the fluid. Increased temperature helps in increasing Brownian motion of nanoparticles assisting thermal conductivity to increase.

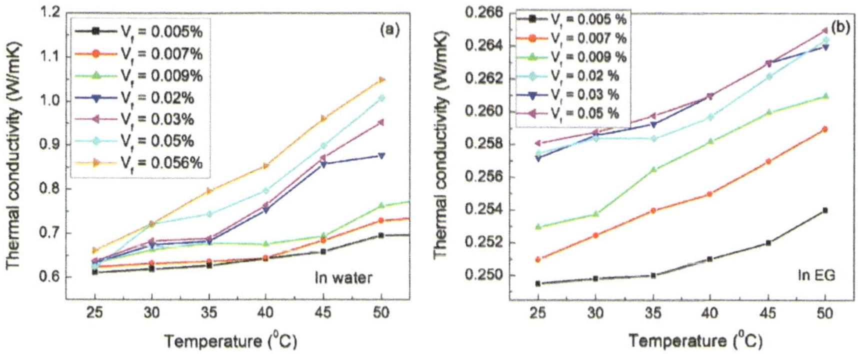

Fig. 5.10 Thermal conductivity of graphene water and EG based nanofluid at various concentration and temperature [11]

Preparation of TiO_2/H_2O

TiO_2/H_2O nanofluid have superior thermal properties and have been extensively used in various applications such as solar energy systems, heat sinks, electronic devices, nucleate pool boiling and heat exchangers. Synthesis of TiO_2 nanoparticles is a key challenge. There are different methods that can be utilized to prepare TiO_2 nanoparticles such as Sol gel method and hydrothermal synthesis. Sol gel method has the advantage of producing nanoparticles with high surface area which is an important parameter for heat transfer application. Sharma et al. [12] prepared TiO_2 nanocrystalline powder using sol gel method from titanium tetra-isopropoxide (as precursor material). Precursor was added into isopropanol and deionized water solution at 80 °C temperature. Solution was then stirred for 6 h in a round flask containing titanium powder, then prepared sol gel was heated for 1 h at 300 °C to obtain nanocrystals of TiO_2.

Temperature is an important parameter in sol gel method during the preparation of TiO_2 nanoparticles. Structure of TiO_2 was changed to rutile from anatase during calcination process when temperature was increased. Size of nanocrystalline was also increased due to change in temperature during calcination process [13]. Sol gel method is explained in Fig. 5.11.

Duangthongsuk et al. [15] dispersed TiO_2 nanoparticles in water by ultrasonication and addition of surfactant (CTAB with 0.01% concentration). Ultrasonication was done for 3 to 4 h for better dispersibility. TEM showed a small agglomeration of nanoparticles after 3 h of sonication. Barzegarian et al. [16] mixed distilled water and CTAB surfactant. In second step, nanoparticles of TiO_2 were added in it and dispersed for 30 min in ultrasonic homogenizer. Ahmed et al. [17] prepared TiO_2/H_2O nanofluid with the help of Triton X-100 surfactant. Triton X-100 helped in decreasing surface tension and contact angle of nanoparticles which allow them to disperse in base fluid with high uniformity allowing highly stable nanofluid.

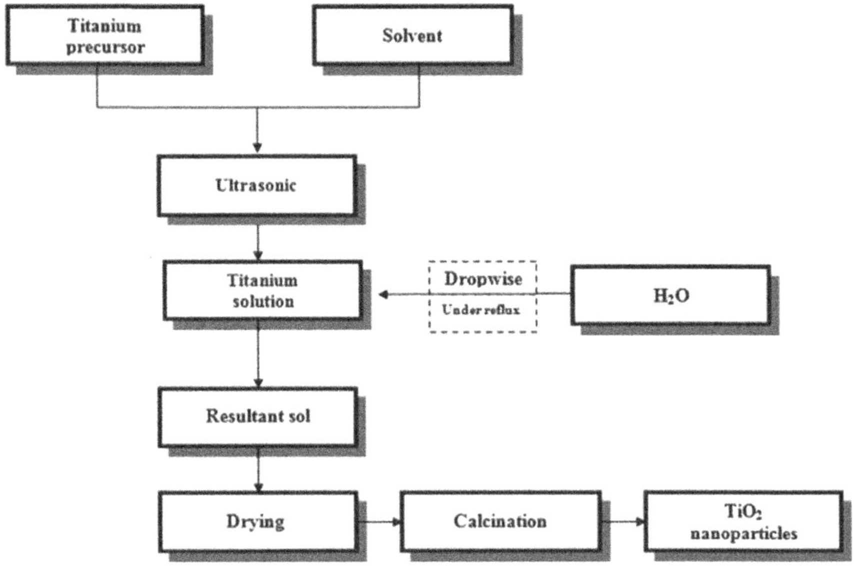

Fig. 5.11 Steps involved in Sol gel method for preparation of TiO_2 nanoparticles [14]

Murshed et al. [18] used cylindrical and spherical nanoparticles to prepare TiO_2/H_2O nanofluid. Nanofluid was prepared by dispersing the TiO_2/H_2O nanoparticles in ultrasonic dismembrator for around 8 to 10 h. In another study, it was found that increase in ultrasonication time will result in breaking of nanoparticle into smaller sizes, decrement in particle size becomes constant after specific ultrasonication time [19] (Fig. 5.12).

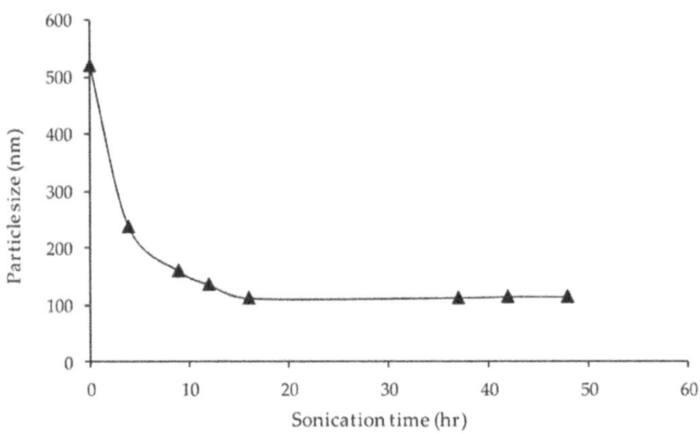

Fig. 5.12 TiO_2 particle size variation with sonication time [19]

Fig. 5.13 Biological synthesis of Ag/EG-H_2O nanofluid [21]

Preparation of Ag/H_2O

Godson et al. [20] used ultra vibrator with frequency up to 100 Hz. Silver/H_2O nanofluid was prepared by using poly vinyl pyrrolidine as surfactant. Ultrasonication was done for 30 min continuously. SEM showed well dispersion of nanoparticles. Zeta potential was used to ensure nanofluid stability. Results showed better stability of fluid due to greater electronegative values of nanoparticles which resulted in greater stability of Ag/H_2O nanofluid.

Sarafraz et al. [21] used biological method to prepare Ag/EG-H_2O nanofluid. Steps involved in preparation are given in Fig. 5.13. Stability of nanofluid was different at different concentrations. Maximum stability achieved was 12 days at 1 vol.%, while 7- and 9-days stability was achieved for 0.1 and 0.5 vol.% concentration, respectively.

Salehi et al. [22] prepared Ag/DI-H_2O using single step method from silver nitrate. Polyvinylpyrrolidone (PVP) was used as a surfactant. Steps involved in preparing nanofluid is given in Fig. 5.14.

Reaction involved in Ag/H_2O nanofluid are:

$$N_2H_4 + AgNO_3 + e^- \rightarrow Ag + N_2 + 2H_2 + NO_3^-$$

$$NaBH_4 + AgNO_3 + 4H_2O \rightarrow B(OH)_4 + Ag + NaNO_3 + 4H_2$$

Preparation of CNT/H_2O

Carbon nanotubes are not soluble in water due to hydrophobic in nature. To obtain a fully dispersed and stable nanofluid (water as base fluid), it is necessary to remove amorphous carbon and residual metal particles by treating it with O_2 functional

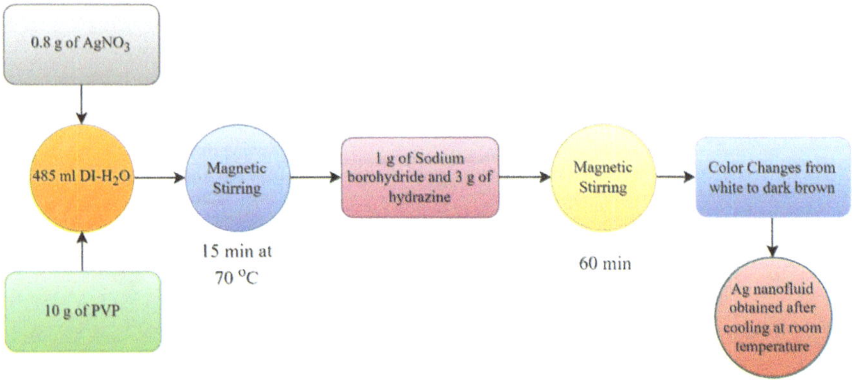

Fig. 5.14 Preparation of Ag/H_2O nanofluid using single step method

groups. Oxidative treatment with sulfuric acid or nitric acid can help in reducing amorphous carbon present on surface of CNT [23]. Another way of preparing a stable nanofluid is the use of suitable surfactant. Yousefi et al. [24] found Triton X-100 to be the best surfactant for CNT/H_2O nanofluid compared to SDS, Tween 80, Tween 20. It is due to the reason that Triton X-100 have benzene ring, molecules with benzene ring structure have the capacity to adsorb more effectively to the graphitic surface because of $\pi - \pi$ stacking type interaction. Optimum amount of Triton X-100 required is less compared to other surfactants for preparing stable CNT/H_2O nanofluid (Fig. 5.15).

Lotfi et al. [25] analyzed that thermal conductivity of CNT/H_2O nanofluid was reduced by addition of surfactant like SDS, Gum Arabia and sodium laurate. So, functionalization of CNT was done instead of adding surfactant. pH of system was

Fig. 5.15 MWCNT/H_2O nanofluid **a** without Triton X-100 **b** with Triton X-100 [24]

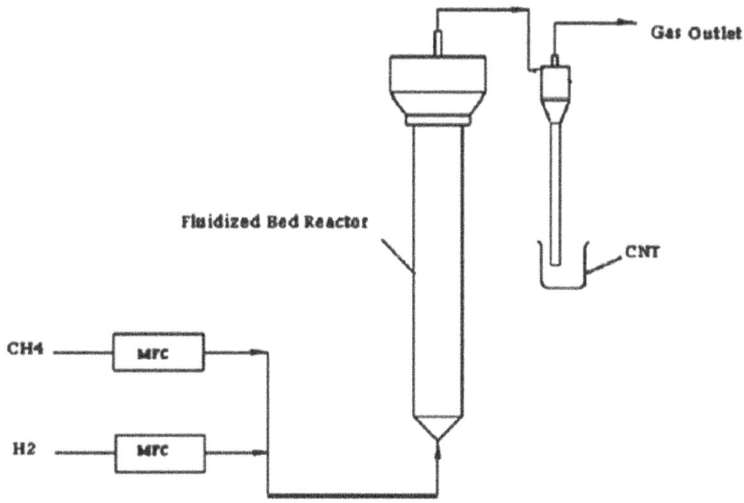

Fig. 5.16 Production of CNT experimentally [25]

controlled by potassium hydroxide. Magnetic stirring was done for 3 h at 85 °C after ultrasonication to prepare a stable MWCNT/H_2O nanofluid (Fig. 5.16).

Goodarzi et al. [26] performed steps given in Fig. 5.17 to functionalize MWCNT with silver and cysteine. Sonication with cysteine was done for 24 h to achieve higher dispersibility of MWCNT in pure H_2O. Poongavanam et al. [27] used tungsten carbide ball milling for 1 h in order to break the agglomeration MWCNT nanoparticles under dry condition. Concentration of surfactant gum Arabic used was 0.25% to prepare a stable MWCNT/H_2O nanofluid.

Preparation of Fe_2O_3/H_2O

Shahrul et al. [28] suggested that 90 min or above period sonication was must to produce a stable nanofluid. Sample vaporization was avoided by applying refrigerated circulating water bath to reduce temperature during sonication. Fe_2O_3 nanoparticles were obtained by applying the chemicals given in Fig. 5.18. Nanoparticles were washed with ethanol and water for 3 times and dried at 70 °C. Dried nanoparticles were calcined at 500 °C in air for 1 to 2 h [29].

Preparation of Al_2O_3/H_2O

Barzegarian et al. [30] used SBDS surfactant in order to reduce adhesion of nanofluid to prepare a stable and well suspended γ-Al_2O_3/H_2O nanofluid. Solution of water, γ-Al_2O_3 and surfactant was homogenized using ultrasonic homogenizer for 60 min. Pandey et al. [31] prepared Al_2O_3/H_2O nanofluid by mixing desired concentration and amount of Al_2O_3 nanoparticles and H_2O in a SKL650-IIN ultrasonic processor for 8 to 16 h. No sedimentation was found after 24 h of preparation. Huang et al. [32] prepared Al_2O_3/H_2O and MWCNT/H_2O nanofluid. Al_2O_3/H_2O nanofluid was prepared without any surfactant and it was stirred for 30 min with ultrasonication

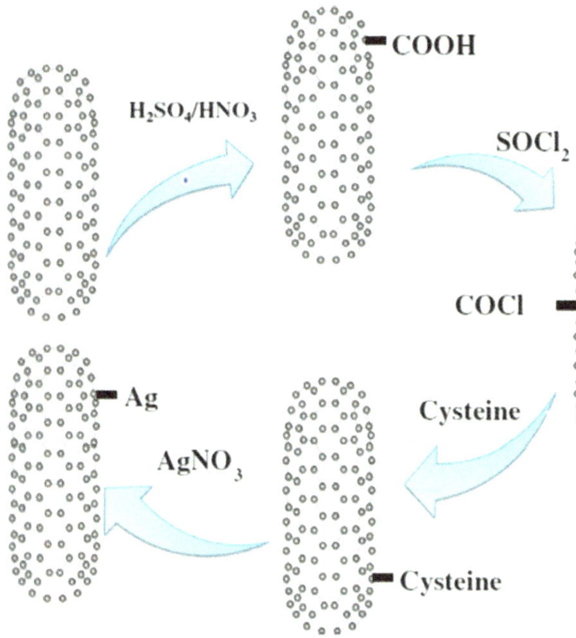

Fig. 5.17 Steps for functionalization of MWCNT with Ag and cysteine [26]

Fig. 5.18 Synthesis of Fe_2O_3 nanoparticles

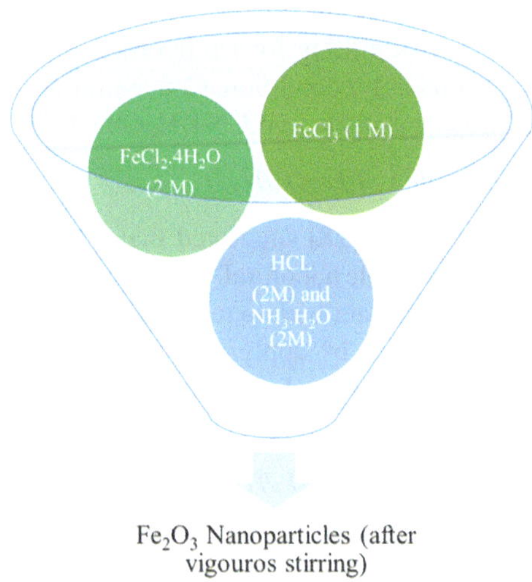

done for 4 h. While, MWCNT/H_2O was made stable by addition of surfactant, it was stirred for 5 min with ultrasonication period of 1 h. Al_2O_3/H_2O was stable for 2 weeks while MWCNT/H_2O was stable for 1 month.

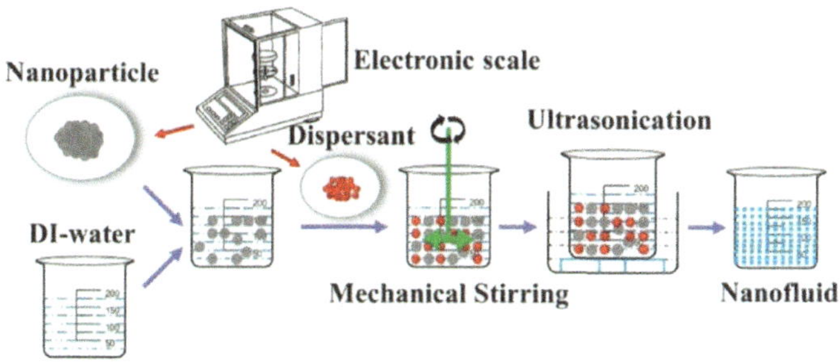

Fig. 5.19 Steps for nanofluid preparation [35]

TiO_2 and Al_2O_3 nano powder were synthesized by using HEBM technique at 300 rpm. Ball to powder ratio used was 10:1. Micro powder was milled for 15 h with 1 h milling time in tungsten carbide jar having tungsten carbide balls. Overheating was avoided by taking a 30 min gap [33]. Tiwari et al. [34] used magnetic stirring along with ultrasonication (Toshiba ultrasonic processor, India) and addition of surfactant (0.01% concentration of CTAB) for preparing of CeO_2/H_2O, Al_2O_3/H_2O, TiO_2/H_2O, SiO_2 /H_2O at different concentrations. No sedimentation was found even at small flow rate during experimentation. Zheng et al. [35] prepared Al_2O_3/H_2O nanofluid in the following steps given in Fig. 5.19 (Table 5.1).

Zheng et al. [35] discussed the stability of nanofluids mentioned in Table 5.1 in terms of zeta potential. Zeta potential index is used to characterize stability of dispersed colloidal nanofluids. Greater the value of zeta potential, greater will be the stability of nanofluid. Generally, for a stable nanofluid, zeta potential is above 25. In another study, Al_2O_3/H_2O nanofluid was prepared by applying 3 h of mechanical stirring and then several hours of ultrasonic vibration (high intensity) [36] (Figs. 5.20 and 5.21).

Preparation of CuO/H_2O

Zhu et al. [37] prepared CuO/H_2O nanofluid using wet chemical method. Steps involved in preparing nanofluid is given in Fig. 5.22.

Khoshvaght-Aliabadi et al. [38] used one step method (Electro exploded wire technique (EEW)) to form stable nanofluid of Cu/H_2O. Nanofluid formed was stable for one week. New correlation for thermal conductivity and viscosity was proposed given in Table A.1. Zamzamian et al. [39] synthesized Al_2O_3/EG and CuO/EG nanofluid. SDS surfactant was used for preparation of CuO/EG nanofluid while no surfactant was used for Al_2O_3/EG. 1-h magnetic stirring and 2 h ultrasonication was done to prepare stable nanofluids. Kim et al. [40] prepared Al_2O_3/H_2O nanofluid by using two step method and amorphous carbonic nanofluid by one step method using electrochemical and sonochemical oxidation.

Pantzali et al. [41] prepared different nanofluids (Al_2O_3, CuO, TiO_2) by mechanical stirring and found that CuO/H_2O had higher thermal conductivity than other nanofluids used. Stability of nanofluid was dependent on size, volume fraction, type, and shape of nanoparticle. An increment in 10% of thermal conductivity was found

Table 5.1 Nanofluid preparation details [35]

Nanofluid	Particle morphology	Size	Mechanical stirring time (min)	Ultrasonication time (hours)	Dispersant	Particle mass: dispersant mass	Zeta potential (stability index) (mV)
Al_2O_3/H_2O	Spherical	30	30	1	Sodium hexametaphosphate	4:1	−36.7
SiC/H_2O	Almost Spherical	40	40	2	CTAB	5:3	−26.8
CuO/H_2O	Almost Spherical	30	30	1.5	Sodium hexametaphosphate	4:1	−34.5
Fe_3O_4/H_2O	Spherical	25	30	1	Sodium citrate	5:1	−29.9

$\phi_1 = 1.2$ vol.% $\phi_2 = 1.9$ vol.% $\phi_3 = 2.6$ vol.%

Fig. 5.20 Al_2O_3/H_2O nanofluid at different volume concentrations [36]

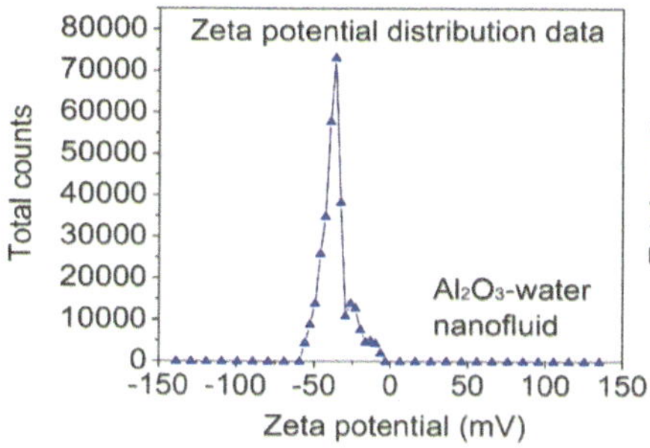

(a) Al_2O_3-water nanofluid

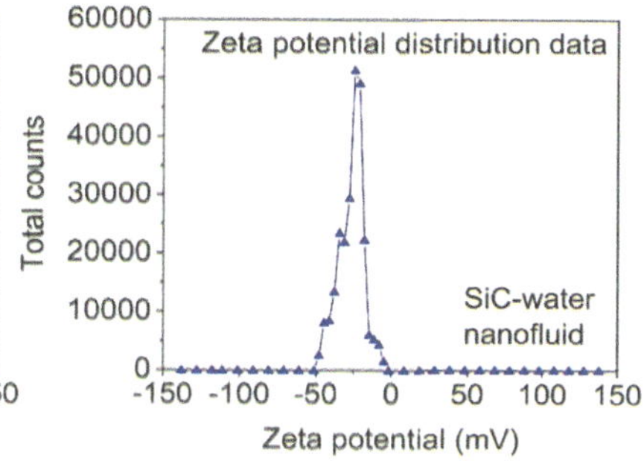

(b) SiC-water nanofluid

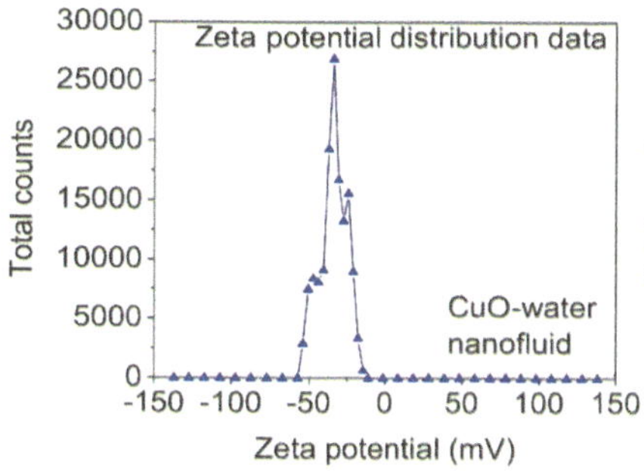

(c) CuO-water nanofluid

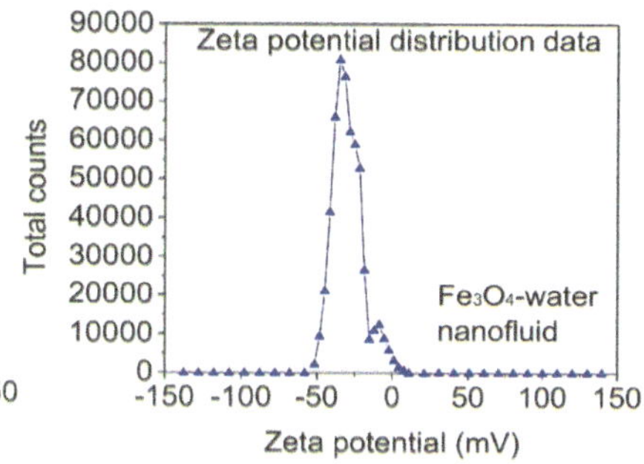

(d) Fe_3O_4-water nanofluid

Fig. 5.21 Zeta potential distribution of nanofluids [35]

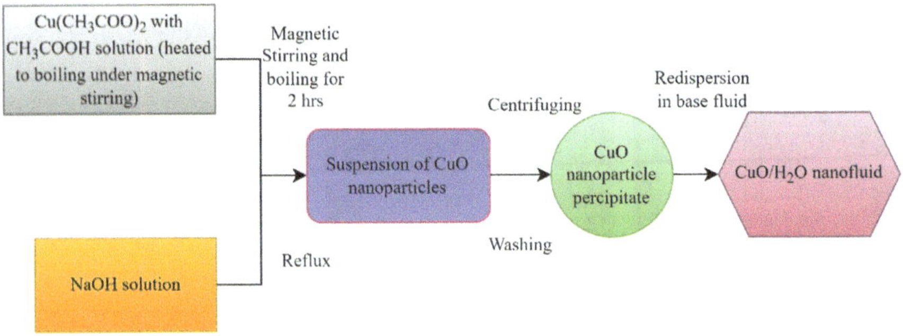

Fig. 5.22 Preparation of CuO/H_2O nanofluid using wet chemical method

as compared to base fluid water. Pantzali et al. [42] applied different techniques like ultrasonic vibration, surfactant addition, pH control to prepare CuO, Al_2O_3 and CNT nanofluids. CuO/H_2O had more promising performance than other two nanofluids but showed poor stability. Stability of CuO/H_2O nanofluid was increased to 3 weeks by addition of CTAB surfactant. Mare et al. [43] measured nanoparticle size using dynamic light scattering technique. DLS was performed with help of Zetasizer Nano S. Karami et al. [44] prepared CuO/H_2O-EG nanofluid using PVP as surfactant. Ultrasonication was done for 60 min with 2 min break after every 15 min to avoid overheating, this provides enough time to nanoparticle to dissipate energy within the fluid. Concentration of PVP was used around 0.25:1 (PVP:CuO).

Shah et al. [45] prepared CuO nanoparticles through precipitation method with and without polyvinyl alcohol (PVA) surfactant. Steps involved in preparing CuO/H_2O-EG nanofluid are given in Fig. 5.23.

From Fig. 5.24, without PVA, prepared CuO nanoparticles were elongated in one direction and were porous while PVA prepared CuO nanoparticles helped in breaking agglomeration of nanoparticles and decreased size.

Preparation of TiN/H_2O

TiN/H_2O nanofluid was prepared through magnetic stirrer. To minimize the surface tension and reduced sedimentation NaOH was used as a surfactant. For 0.5% volume concentration of nanofluid, 10 ml of NaOH was mixed with 2 L of H_2O through magnetic stirrer for 15 min and then 10 g of TiN_2 powder was mixed with the solution for 30 min. Thermal conductivity, density and specific heat at various concentrations are shown in [46] (Table 5.2).

Preparation of SiC/H_2O

SiC/H_2O nanofluid was prepared by mixing SiC nanoparticles of 50 nm size with water with the help of magnetic stirrer for 30 min and then ultrasonication was done for 60 min to break any agglomeration present within the nanoparticles. For stability of nanofluid, SDS ($NaC_{12}H_{25}SO_4$) was used as a surfactant (around 3% of total weight) [47].

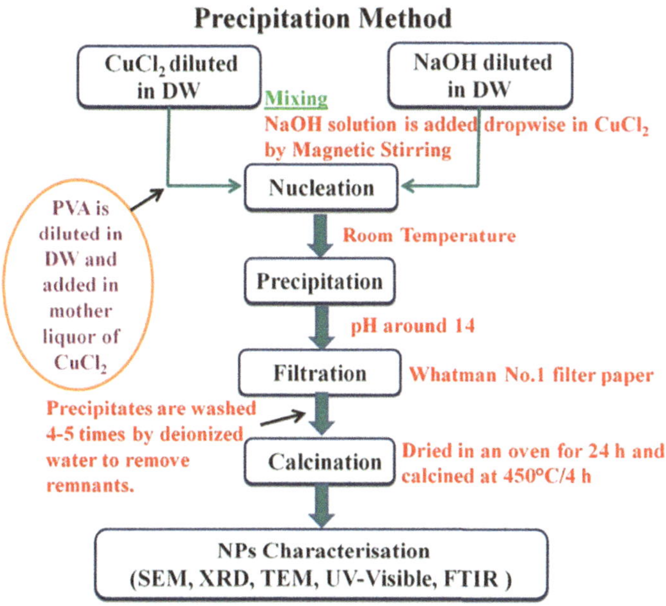

Fig. 5.23 Schematic of CuO nanoparticle preparation with and without PVA

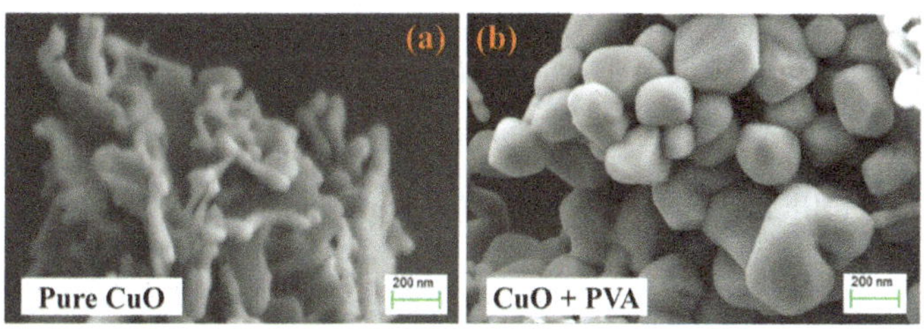

Fig. 5.24 SEM image of **a** Pure CuO **b** CuO coated with polyvinyl alcohol

Table 5.2 Thermal properties of TiN/H_2O nanofluid

Concentration	Thermal conductivity	Density	Specific heat
TiN 0.5	0.7836	1404.15	2.772
TiN 1	1.04597	1793.72	2.011
TiN 2	1.832	2612.04	1.153

5.2 Hybrid Nanofluid Preparation

Mono nanofluids are much better than conventional base fluids for thermal application but they possess one basic issue, either mono nanofluid have good rheological properties or better thermal network. Mono nanofluids unfortunately do not have all favorable properties required for certain applications. For example, metal oxides possess good stability and chemical inertness but they have low thermal conductivity compared to superior nanoparticles (in terms of thermal properties) such as graphene and MWCNT. In recent years, researchers have turned their attention toward hybrid nanofluids which possess both thermal and rheological properties. Hybrid nanofluid is the mixture of two different types or similar types of nanoparticles. A brief overview of synthesis and preparation of some hybrid nanofluid is given in this chapter:

Preparation of Fe_2O_3-CNT hybrid nanofluid

Aghabozorg et al. [48] synthesized hybrid Fe_2O_3-CNT by using ferrocene in toluene at 2.5% concentration. Temperature of furnace was set at 1000 °C and nitrogen was sprayed for 10 min to make N_2 atmosphere instead of air. Mixture of toluene and ferrocene was passed through nozzle and then mixture of toluene/ferrocene/nitrogen was heated at 1000 °C temperature until deposition of Fe_2O_3-CNT. Deposited material was cooled by flowing N_2 at ambient temperature. After collecting Fe_2O_3-CNT, nanoparticles were sonicated with water for 3 h to form a stable nanofluid for different concentrations (0.1 and 0.2 wt%).

Hybrid MWCNT coated with Fe_3O_4/H_2O nanofluid was prepared through chemical reduction process. MWCNT was treated with strong acids (HCl, $FeCl_24H_2O$, $FeCl_36H_2O$) and placing carboxyl group between MWCNT and Fe_3O_4. 500 ml of HNO_3 and 13 M HCl was poured into a 2-L beaker containing MWCNT, mixture was stirred using magnetic stirrer at 300 rpm and 60 °C for 72 h. MWCNT coated with Fe_3O_4 then cleaned with H_2O and dehydrated for 24 h at 90 °C. MWCNT-Fe_3O_4 was then agitated with water for 2 h to obtain stable nanofluid [49].

Preparation of Al_2O_3-MWCNT/H_2O nanofluid

Huang et al. [50] prepared hybrid Al_2O_3-MWCNT/H_2O nanofluid by making Al_2O_3/H_2O and MWCNT/H_2O at 1.89 vol.% and 0.0111 vol.% concentration, respectively. Both nanofluid were mixed together at ratio of 1:2.5 and sonicated for 3 h.

Preparation of MWCNT-CuO

MWCNT-CuO/H_2O nanofluid was prepared using non-ionic surfactant because cationic and anionic surfactant are foamier and caused reduction in heat transfer properties resulted in poor heat transfer process. Weight fraction of surfactant was determined to be 1:1 (surfactant-nanoparticle) and 0.5:1. Surfactant was stirred with water first at maximum speed for 1 h then nanoparticles were added slowly, and solution was again stirred for 1 h. And then ultrasonication was done for 3 h [51].

Synthesis of MWCNT-TiO_2 nanoparticles

Figure 5.25 showed steps to prepare MWCNT-TiO_2 nanoparticles provided by [52].

Synthesis of Al_2O_3–TiO_2 nanocomposites

Figure 5.26 showed steps and compounds involved in preparation of Al_2O_3–TiO_2 nanoparticles given by [53]. Hybrid nanofluid was then prepared by applying

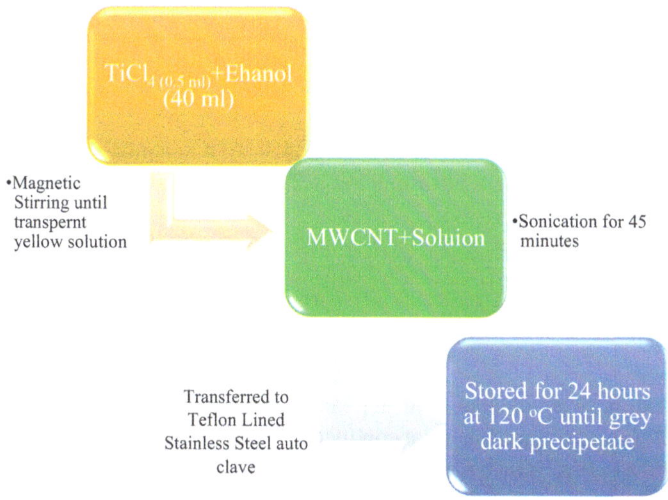

Fig. 5.25 Preparation of MWCNT-TiO_2 nanoparticle

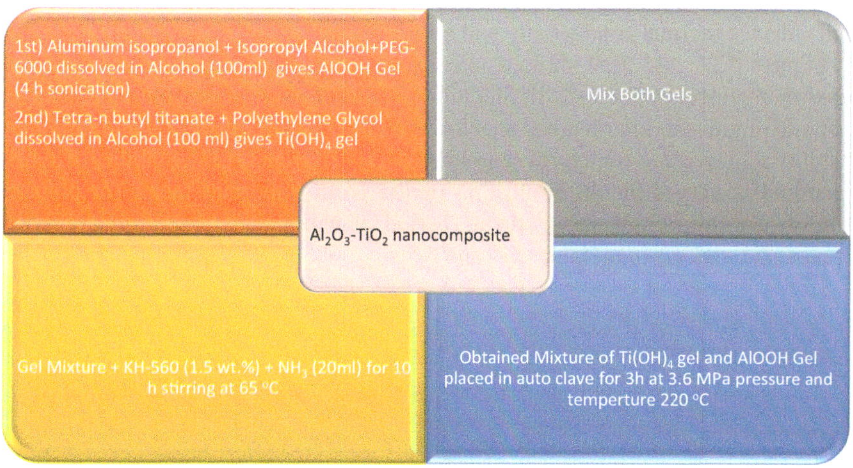

Fig. 5.26 Preparation of Al_2O_3–TiO_2 nanoparticles

ultrasonic vibration for 5 h at 100 W. Surfactant was not utilized because it caused variation in thermal properties of nanofluid. pH of nanofluid was maintained between 5.62 and 7.

Preparation of Al_2O_3-nanoparticles (MWCNT, SiC, AlN, MgO, CuO)/H_2O

Bhattad et al. [54] prepared hybrid nanofluid considering Al_2O_3 nanoparticle as the base nanoparticle and changing the second one. CTAB and SFS were used as surfactant to avoid agglomeration and have better dispersion of hybrid nanoparticles in base fluid. Both surfactants acted differently with different nanoparticles due to cationic nature of CTAB and anionic nature of SDS. It was suggested that CTAB is best for SiC, AlN and MWCNT while SDS was better for MgO and CuO. Hybrid nanofluid was prepared with concentration of 4:1 (Al_2O_3-nanoparticles) with 1 h of mechanical stirring and 4 h of ultrasonication. Thermal properties of hybrid nanofluid given in Table 5.3.

5.3 Discussion

Thermal conductivity of nanofluid depends on temperature and concentration of nanoparticles in nanofluid. Thermal conductivity increases with increase in temperature, as shown in Fig. 5.27, because at higher temperature, agglomeration of nanoparticles break and uniform dispersion takes place. Also, Brownian motion of nanoparticles increase at higher temperature as compared to lower temperature. Concentration of nanoparticles also had a direct relation with thermal conductivity because increase in concentration increases the number of heat carrying nanoparticles which will enhance thermal conductivity and ultimately increases heat transfer rate.

Figures 5.28 and 5.29 showed maximum stability of nanofluid achieved and maximum concentration used in literature. Results from different literature showed that maximum stability obtained was for MWCNT/Solar Glycol [27] and minimum stability achieved was for hybrid nanofluid Al_2O_3–TiO_2/H_2O [33]. Stability of hybrid nanofluid is very difficult. Sonication period is an important factor for preparation of nanofluid. Sonication, stirring period of nanofluid and thermal properties of nanoparticles given in Tables 5.4 and 5.5.

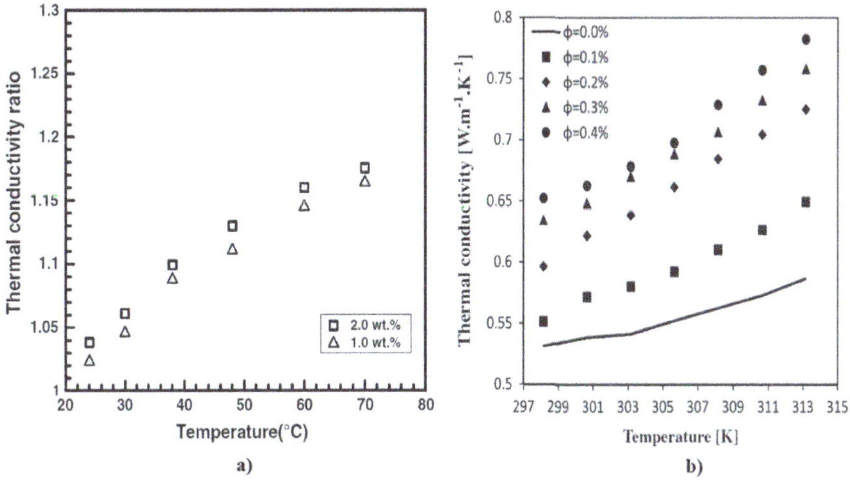

Fig. 5.27 Variation of thermal conductivity with temperature and concentration **a** [70] **b** [38]

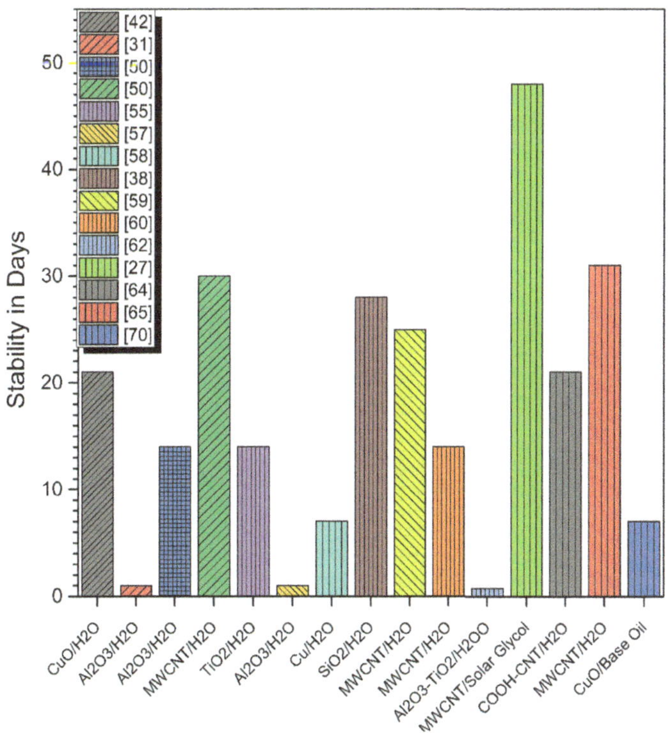

Fig. 5.28 Stability of various nanofluids

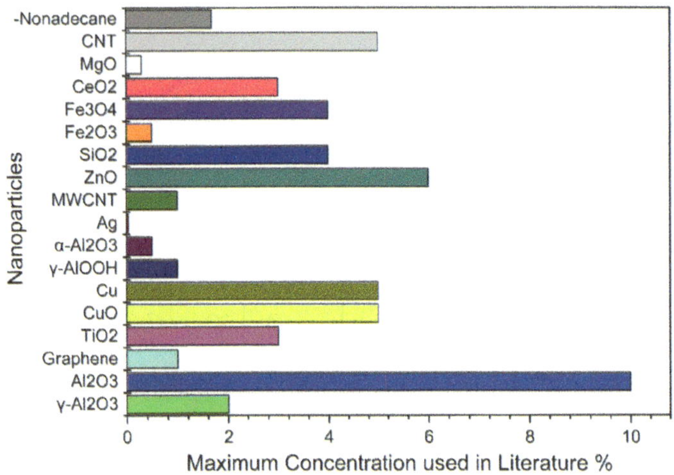

Fig. 5.29 Maximum concentration used in literature

Table 5.3 Hybrid nanofluid thermal properties [54]

Nanofluid (water base fluid)	Thermal conductivity (W/mK)	Specific heat (J/kgK)	Density (kg/m^3)	Viscosity (Pas)
Al_2O_3-MWCNT	0.611	4168	997.6	0.0008806
Al_2O_3-CuO	0.5989	4167	1000.1	0.0008798
Al_2O_3-MgO	0.6018	4172	998.4	0.0008752
Al_2O_3-SiC	0.6063	4174	999.2	0.0008732
Al_2O_3-AlN	0.6052	4169	999.3	0.0008742

Table 5.4 Summary of preparation and stability of nanofluids

Refs. No.	Nanofluid	Size nm	Surfactant	Conc.%	Method	Thermal conductivity measurement	Stability period
[20]	Ag/H_2O	54	PVP	0.01–0.04 vol.%	Ultrasonication with surfactant addition		
[29]	Fe_2O_3/EG Fe_2O_3/H_2O	22	Not used	0.01–0.08 vol.%	Sonication for 50 min at 40 kHz and 1200 W	KD-2 Pro	
[6]	Graphene Oxide/H_2O	650–1220		0.01 and 0.1 wt.%	Stirring for 5 h and sonication for 1 h at 130 W and 42 kHz	Transient hot wire method	
[41]	Al_2O_3, CuO, TiO_2/H_2O	30	Not used	2–8 vol.%	Mechanical stirring	Transient hot wire method	
[42]	CuO/H_2O	30	CTAB	4 vol.%	Ultrasonic vibration with surfactant addition		3 weeks
[31]	Al_2O_3/H_2O	40–50	Not used	2–4 vol.%		TPS500 instrument	No sedimentation observed after 24 h
[34]	CeO_2, Al_2O_3, TiO_2, SiO_2/H_2O	30, 45, 10, 10	CTAB	0.5–3 vol.%	Ultrasonication and magnetic stirring with surfactant addition	Transient hot wire (KD-2 analyzer)	No sedimentation during experiment
[26]	MWCNT-Ag/H_2O	10–20	GA	0–1 wt.%	Sonication for 24 h	Transient hot wire (KD-2 analyzer)	

(continued)

Table 5.4 (continued)

Refs. No.	Nanofluid	Size nm	Surfactant	Conc.%	Method	Thermal conductivity measurement	Stability period
[32]	MWCNT, Al_2O_3/DI H_2O	9.5 MWCNT, 40 Al_2O_3	SDBS	2.18–10.36 Al_2O_3, 0.02–0.1 wt.%	Al_2O_3 stirring for 30 min and sonication for 4 h and MWCNT stirring for 5 min and sonication for 1 h		Al_2O_3/H_2O = 2 weeks MWCNT/H_2O = 1 month
[55]	TiO_2/H_2O	15	Not used	0.25–0.8 vol.%	Magnetic stirring for 4 h and ultrasonication for 2 h		No sedimentation after 2 weeks
[56]	ZnO/H_2O		CTAB	0.25–2 vol.%	Ultrasonic vibration for 6 h	KD2 Pro-thermal analyzer	
[57]	MWCNT/H_2O	5–15	GA	0.25–0.55 wt.%	4 h magnetic stirring and then ultrasonication for 2 h		No sedimentation after 2 days
[58]	Al_2O_3/H_2O	47	Not used	1–4 vol.%	Physical vapor synthesis and ultrasonication for 4 h	Transient jot wire method	No precipitation after 24 h
[38]	Cu/H_2O	30–50		0.1–0.4 vol.%	One step method	Transient line source method (KD2 Pro)	1 week
[59]	SiO_2/H_2O		Tetrasodium pyrophosphate	1 and 2 vol.%	Magnetic stirring for 10 min and ultrasonic vibration for 30 min		Up to 4 weeks
[60]	MWCNT/H_2O		Nonylphenolethoxilate	0.5–1.5 vol.%			14 days

(continued)

Table 5.4 (continued)

Refs. No.	Nanofluid	Size nm	Surfactant	Conc.%	Method	Thermal conductivity measurement	Stability period
[61]	γ-Al_2O_3/H_2O	40	Not used	0.78–7.04 wt.%	30 min mechanical stirring and 4 h ultrasonic vibration		
[62]	$MWCNT/H_2O$	3–5 nm	CTAB (0.1%)	0.04–0.25 wt.%	Ultrasonication at 300 W and 40 kHz		25 days
[33]	Al_2O_3–TiO_2/H_2O	30–40	Polyvinyl Alcohol	0.025–0.1 vol.%	Magnetic stirring for 6 h at 1000 rpm than ultrasonication for 2 h		12–18 h
[63]	Fe_3O_4/H_2O		CTAB (1/10th wt. of nanoparticles)	0.005–0.06 vol.%	Stirring for full dispersion continued to 8 h		
[27]	MWCNT/Solar Glycol	20–30	GA	0.2–0.6 vol.%	60 min sonication for 0.2 and 0.4 vol.% and 90 min sonication for 0.6 vol.%		More than 48 days
[64]	$COOH\text{-}CNT/H_2O$	10–20	Nonylphenol ethoxylate (0.1%)	0.1–0.3 wt.%	Ultrasonication for 10 min at 300 W and 40 kHz		21 days
[65]	$MWCNT/H_2O$		CTAB (1:1)	0.04–0.25 wt.%	Ultrasonic bath at 40 kHz and 300 W for 120 min		More than 30 days
[66]	CuO/water	30–50	Not used	> 0.24 vol.%	Vibrated for 10 h		No precipitation after 5 h
[67]	TiO_2/H_2O	30	Not used	0.05–0.25 vol.%	10 min mixing and ultrasonication for 30 min		

(continued)

Table 5.4 (continued)

Refs. No.	Nanofluid	Size nm	Surfactant	Conc.%	Method	Thermal conductivity measurement	Stability period
[68]	Al_2O_3/H_2O	47	SDBS	0.02–0.5 vol.%	Magnetic stirring for continuous 10 h		
[69]	γ-Al_2O_3, TiO_2, CuO/DI H_2O	25 γ-Al_2O_3, 10 TiO_2, 30–50 CuO	CMC	0.2 vol.%	Ultrasonic vibration of mixture then mechanical stirring with addition of CMC		Several days
[70]	CuO/base oil	50	Not used	0.2–2 vol.%	Ultrasonication at 400 W and 24 kHz		Complete sedimentation in a week
[71]	TiO_2/H_2O	15 nm	Not used	0.07–0.21 vol.%	Sonication of well dispersed solution for 3 h at 100 W		More than 3 h
[72]	Al_2O_3/H_2O CuO/H_2O	20 and 40 nm	Not used	0.5–1.5 vol.%	1 h magnetic stirring followed by 5 h of ultrasonication		No sedimentation after 3 weeks

Table 5.5 Physical properties of nanoparticles given in literature

Refs.	Nanoparticle	Size nm	Density kg/m^3	Thermal conductivity W/m^{2}K	Specific heat J/kgK
[73]	γ-Al_2O_3	25	3700	46	880
[74]	α-Al_2O_3	50	3880	36	773
[75, 76]	Al_2O_3	50	3970	205	765
[7]	Graphene		2200	5000	790
[77]	γ-AlOOH		3050	30	618.3
[78]	TiO_2	21	4260	11.7	697
[79]	SiO_2		2220	1.38	745
[78]	ZnO	50	7133	111	383
[80]	CuO		6510	18	540
[54]	MWCNT		2600	3000	735
[81]	Cu	25	8940	397.5	385
[82]	AlN		4175	8.4	692
[83]	Fe_3O_4	36	5180	80.4	670
[49]	MWCNT-Fe_3O_4		4845.4	787.27	680.66

References

1. Sandhya, M., et al. (2021). A systematic review on graphene-based nanofluids application in renewable energy systems: Preparation, characterization, and thermophysical properties. *Sustainable Energy Technologies and Assessments, 44*, 101058.
2. Ali, H. M., et al. (2018). Preparation techniques of TiO_2 nanofluids and challenges: A review. *Applied Sciences, 8*(4), 587.
3. Choi, S. U., & Eastman, J. A. (2001). *Enhanced heat transfer using nanofluids*. Google Patents.
4. Sun, B., et al. (2016). Investigation on the flow and convective heat transfer characteristics of nanofluids in the plate heat exchanger *76*, 75–86.
5. Bahiraei, M., & Heshmatian, S. (2019). Graphene family nanofluids: A critical review and future research directions. *Energy Conversion and Management, 196*, 1222–1256.
6. Esfahani, M. R., & Languri, E. M. (2017). Exergy analysis of a shell-and-tube heat exchanger using graphene oxide nanofluids. *83*, 100–106.
7. Ghozatloo, A., et al. (2014). Convective heat transfer enhancement of graphene nanofluids in shell and tube heat exchanger. *53*, 136–141.
8. Fares, M., Mohammad, A.-M., & Mohammed, A.-S. (2020). Heat transfer analysis of a shell and tube heat exchanger operated with graphene nanofluids. *Case Studies in Thermal Engineering, 18*, 100584.
9. Wang, Z., et al. (2018). Experimental comparative evaluation of a graphene nanofluid coolant in miniature plate heat exchanger. *130*, 148–156
10. Akhavan-Zanjani, H., et al. (2016). Experimental investigation of laminar forced convective heat transfer of graphene–water nanofluid inside a circular tube. *100*, 316–323.
11. Baby, T. T., & Ramaprabhu, S. (2010). Investigation of thermal and electrical conductivity of graphene based nanofluids. *Journal of Applied Physics, 108*(12), 124308.
12. Sharma, A., Karn, R., & Pandiyan, S. (2014). Synthesis of TiO_2 nanoparticles by sol-gel method and their characterization. *Journal of Basic and Applied Engineering Research, 1*(9), 1–5.

13. Cenovar, A., et al. (2012). Preparation of nano-crystalline TiO_2 by sol-gel method using titanium tetraisopropoxide (TTIP) as a precursor. *Advances in Natural Science: Theory & Applications.*
14. Behnajady, M., et al. (2011). Investigation of the effect of sol–gel synthesis variables on structural and photocatalytic properties of TiO_2 nanoparticles. *Desalination, 278*(1–3), 10–17.
15. Duangthongsuk, W., Wongwises, S. (2009). Heat transfer enhancement and pressure drop characteristics of TiO_2–water nanofluid in a double-tube counter flow heat exchanger. *52*(7–8), 2059–2067.
16. Barzegarian, R., et al. (2016). Experimental investigation on heat transfer characteristics and pressure drop of BPHE (brazed plate heat exchanger) using TiO2–water nanofluid. *74*, 11–18.
17. Ahmed, S. A., et al. (2018). Improving car radiator performance by using TiO_2-water nanofluid. *Engineering Science and Technology, an International Journal, 21*(5), 996–1005.
18. Murshed, S., Leong, K., & Yang, C. (2005). Enhanced thermal conductivity of TiO_2—water based nanofluids. *International Journal of Thermal Sciences, 44*(4), 367–373.
19. Palabiyik, I., et al. (2011). Dispersion stability and thermal conductivity of propylene glycol-based nanofluids. *Journal of Nanoparticle Research, 13*(10), 5049–5055.
20. Godson, L., et al. (2014). Heat transfer characteristics of silver/water nanofluids in a shell and tube heat exchanger. *14*(3), 489–496.
21. Sarafraz, M., & Hormozi, F. (2015). Intensification of forced convection heat transfer using biological nanofluid in a double-pipe heat exchanger. *66*, 279–289.
22. Salehi, J., Heyhat, M., & Rajabpour, A. (2013). Enhancement of thermal conductivity of silver nanofluid synthesized by a one-step method with the effect of polyvinylpyrrolidone on thermal behavior. *Applied Physics Letters, 102*(23), 231907.
23. Afzal, A., et al. (2018). Heat transfer characteristics of MWCNT nanofluid in rectangular mini channels. *International Journal of Heat and Technology, 36*(1), 222–228.
24. Yousefi, T., et al. (2012). An experimental investigation on the effect of MWCNT-H_2O nanofluid on the efficiency of flat-plate solar collectors. *Experimental Thermal and Fluid Science, 39*, 207–212.
25. Lotfi, R., et al. (2012). Experimental study on the heat transfer enhancement of MWNT-water nanofluid in a shell and tube heat exchanger. *39*(1), 108–111
26. Goodarzi, M., et al. (2015). Investigation of heat transfer and pressure drop of a counter flow corrugated plate heat exchanger using MWCNT based nanofluids. *66*, 172–179.
27. Poongavanam, G. K., et al. (2019). Experimental investigation on heat transfer and pressure drop of MWCNT-Solar glycol based nanofluids in shot peened double pipe heat exchanger. *345*, 815–824.
28. Shahrul, I., et al. (2016). Experimental investigation on Al_2O_3–W, SiO_2–W and ZnO–W nanofluids and their application in a shell and tube heat exchanger. *97*, 547–558.
29. Kumar, N., & Sonawane, S. S. (2016). Experimental study of Fe_2O_3/water and Fe2O3/ethylene glycol nanofluid heat transfer enhancement in a shell and tube heat exchanger. *78*, 277–284
30. Barzegarian, R., et al. (2017). Thermal performance augmentation using water based Al2O3-gamma nanofluid in a horizontal shell and tube heat exchanger under forced circulation. *86*, 52–59.
31. Pandey, S. D., & Nema, V. K. (2012). Experimental analysis of heat transfer and friction factor of nanofluid as a coolant in a corrugated plate heat exchanger. *38*, 248–256.
32. Huang, D., et al. (2015). Pressure drop and convective heat transfer of Al_2O_3/water and MWCNT/water nanofluids in a chevron plate heat exchanger. *89*, 620–626.
33. Mund, A., et al. (2019). Experimental and numerical study of heat transfer in double-pipe heat exchanger using Al_2O_3, and TiO_2 water nanofluid. In *Advances in fluid and thermal engineering*. Springer, p 531–540.
34. Tiwari, A. K., et al. (2013). Performance comparison of the plate heat exchanger using different nanofluids. *49*, 141–151.
35. Zheng, D., et al. (2020). Performance analysis of a plate heat exchanger using various nanofluids. *International Journal of Heat and Mass Transfer, 158*, 119993.

36. Attalla, M., & Maghrabie, H. M. (2020). An experimental study on heat transfer and fluid flow of rough plate heat exchanger using Al_2O_3/water nanofluid. *Experimental Heat Transfer, 33*(3), 261–281.
37. Zhu, H., et al. (2011). Preparation and thermal conductivity of CuO nanofluid via a wet chemical method. *Nanoscale Research Letters, 6*(1), 1–6.
38. Khoshvaght-Aliabadi, M., et al. (2014). Experimental analysis of thermal–hydraulic performance of copper–water nanofluid flow in different plate-fin channels. *52*, 248–258.
39. Zamzamian, A., et al. (2011). Experimental investigation of forced convective heat transfer coefficient in nanofluids of Al_2O_3/EG and CuO/EG in a double pipe and plate heat exchangers under turbulent flow. *35*(3), 495–502.
40. Kim, D., et al. (2009). Convective heat transfer characteristics of nanofluids under laminar and turbulent flow conditions. *9*(2), e119–e123
41. Pantzali, M., et al. (2009). Effect of nanofluids on the performance of a miniature plate heat exchanger with modulated surface. *30*(4), 691–699.
42. Pantzali, M., Mouza, A., & Paras, S. V. (2009). Investigating the efficacy of nanofluids as coolants in plate heat exchangers (PHE). *64*(14), 3290–3300.
43. Maré, T., et al. (2011). Comparison of the thermal performances of two nanofluids at low temperature in a plate heat exchanger. *35*(8), 1535–1543.
44. Karami, M., et al. (2015). Experimental investigation of CuO nanofluid-based direct absorption solar collector for residential applications. *Renewable and Sustainable Energy Reviews, 52*, 793–801.
45. Shah, J., et al. (2019). Surfactant prevented growth and enhanced thermophysical properties of CuO nanofluid. *Journal of Molecular Liquids, 283*, 550–557.
46. Arulkumar, S., & Mathanraj, V. (2020). Heat transfer enhancement with titanium nitride nanofluid in a shell and tube heat exchanger. *IOP Conference Series: Materials Science and Engineering*.
47. Karimi, S., et al. (2020). Experimental investigation of convective heat transfer and pressure drop of SiC/water nanofluid in a shell and tube heat exchanger. *Heat and Mass Transfer, 56*(8), 2325–2331.
48. Aghabozorg, M. H., et al. (2016). Experimental investigation of heat transfer enhancement of Fe_2O_3-CNT/water magnetic nanofluids under laminar, transient and turbulent flow inside a horizontal shell and tube heat exchanger. 72, 182–189.
49. Alklaibi, A., Sundar, L. S., & Mouli, K. V. C. (2022). Experimental investigation on the performance of hybrid Fe_3O_4 coated MWCNT/Water nanofluid as a coolant of a Plate heat exchanger. *International Journal of Thermal Sciences, 171*, 107249.
50. Huang, D., et al. (2016). Effects of hybrid nanofluid mixture in plate heat exchangers. *72*, 190–196.
51. Fazeli, I., Emami, M. R. S., & Rashidi, A. (2021). Investigation and optimization of the behavior of heat transfer and flow of MWCNT-CuO hybrid nanofluid in a brazed plate heat exchanger using response surface methodology. *International Communications in Heat and Mass Transfer, 122*, 105175.
52. Safi, M., et al. (2014). Calculation of heat transfer coefficient of MWCNT-TiO_2 nanofluid in plate heat exchanger. *10*(3), 153–162.
53. Maddah, H., et al. (2018). Factorial experimental design for the thermal performance of a double pipe heat exchanger using Al_2O_3–TiO_2 hybrid nanofluid. *97*, 92–102.
54. Bhattad, A., et al. (2018). Discrete phase numerical model and experimental study of hybrid nanofluid heat transfer and pressure drop in plate heat exchanger. *91*, 262–273.
55. Taghizadeh-Tabari, Z., et al. (2016). The study on application of TiO_2/water nanofluid in plate heat exchanger of milk pasteurization industries. *58*, 1318–1326.
56. Kumar, V., et al. (2016). Effect of chevron angle on heat transfer performance in plate heat exchanger using ZnO/water nanofluid. *118*, 142–154.
57. Tabari, Z. T., & Heris, S. Z. (2015). Heat transfer performance of milk pasteurization plate heat exchangers using MWCNT/water nanofluid. *36*(2), 196–204.

58. Kabeel, A. E., El Maaty, T. A., & El Samadony, Y. A. F. (2013). The effect of using nano-particles on corrugated plate heat exchanger performance. *52*(1), 221–229.
59. Stogiannis, I., Mouza, A., & Paras, S. V. (2015). Efficacy of SiO_2 nanofluids in a miniature plate heat exchanger with undulated surface. *92*, 230–238.
60. Sarafraz, M., & Hormozi, F. (2016). Heat transfer, pressure drop and fouling studies of multi-walled carbon nanotube nano-fluids inside a plate heat exchanger. *72*, 1–11.
61. Wu, Z., Wang, L., & Sunden, B. (2013). Pressure drop and convective heat transfer of water and nanofluids in a double-pipe helical heat exchanger. *60*(1–2), 266–274.
62. Moradi, A., et al. (2019). An experimental study on MWCNT–water nanofluids flow and heat transfer in double-pipe heat exchanger using porous media. 1–11.
63. Sundar, L. S., et al. (2019). Heat transfer and effectiveness experimentally-based analysis of wire coil with core-rod inserted in Fe_3O_4/water nanofluid flow in a double pipe U-bend heat exchanger. *134*, 405–419.
64. Sarafraz, M., et al. (2016). Thermal performance of a counter-current double pipe heat exchanger working with COOH–CNT/water nanofluids. *78*, 41–49.
65. Hosseinian, A., et al. (2018). Experimental investigation of surface vibration effects on increasing the stability and heat transfer coefficient of MWCNTs-water nanofluid in a flexible double pipe heat exchanger. *90*, 275–285.
66. Fotukian, S. M., & Esfahany, M. N. (2010). Experimental study of turbulent convective heat transfer and pressure drop of dilute CuO/water nanofluid inside a circular tube. *37*(2), 214–219.
67. Sajadi, A., & Kazemi, M. H. (2011). Investigation of turbulent convective heat transfer and pressure drop of TiO_2/water nanofluid in circular tube. *38*(10), 1474–1478.
68. Sundar, L. S., & Sharma, K. V. (2010). Turbulent heat transfer and friction factor of Al_2O_3 nanofluid in circular tube with twisted tape inserts. *53*(7–8), 1409–1416.
69. Hojjat, M., et al. (2011). Convective heat transfer of non-Newtonian nanofluids through a uniformly heated circular tube. *50*(4), 525–531.
70. Saeedinia, M., et al. (2012). Thermal and rheological characteristics of CuO–base oil nanofluid flow inside a circular tube. *39*(1), 152–159.
71. Eiamsa-Ard, S., et al. (2015). Heat transfer enhancement of TiO_2/water nanofluid in a heat exchanger tube equipped with overlapped dual twisted-tapes. *18*(3), 336–350.
72. Tong, Y., et al. (2019). Energy and exergy comparison of a flat-plate solar collector using water, Al_2O_3 nanofluid, and CuO nanofluid. *Applied Thermal Engineering, 159*, 113959.
73. Farajollahi, B., et al. (2010). Heat transfer of nanofluids in a shell and tube heat exchanger. *53*(1–3), 12–17.
74. Khoshvaght-Aliabadi, M., et al. (2016). Thermal–hydraulic performance of wavy plate-fin heat exchanger using passive techniques: perforations, winglets, and nanofluids. *78*, 231–240.
75. Nallusamy, S. (2017). Characterization of Al_2O_3/water nanofluid through shell and tube heat exchangers over parallel and counter flow. *Journal of Nano Research.*
76. Nallusamy, S., & Manikanda Prabu, N. (2017). Heat transfer enhancement analysis of Al_2O_3-water nanofluid through parallel and counter flow in shell and tube heat exchangers. *16*(05n06), 1750020.
77. Elias, M., et al. (2014). Effect of different nanoparticle shapes on shell and tube heat exchanger using different baffle angles and operated with nanofluid. *70*, 289–297.
78. Shahrul, I., et al. (2014). Effectiveness study of a shell and tube heat exchanger operated with nanofluids at different mass flow rates. *65*(7), 699–713.
79. Javadi, F. S., et al. (2013). The effects of nanofluid on thermophysical properties and heat transfer characteristics of a plate heat exchanger. *44*, 58–63.
80. Saeedan, M., et al. (2016). CFD Investigation and neutral network modeling of heat transfer and pressure drop of nanofluids in double pipe helically baffled heat exchanger with a 3-D fined tube. *100*, 721–729.
81. Heris, S. Z., Esfahany, M. N., & Etemad, G. (2007). Numerical investigation of nanofluid laminar convective heat transfer through a circular tube. *Numerical Heat Transfer, Part A: Applications, 52*(11), 1043–1058.

82. Hussein, A. M. (2017). Thermal performance and thermal properties of hybrid nanofluid laminar flow in a double pipe heat exchanger. *88*, 37–45.
83. Kumar, N. R., et al. (2017). Heat transfer, friction factor and effectiveness of Fe_3O_4 nanofluid flow in an inner tube of double pipe U-bend heat exchanger with and without longitudinal strip inserts. *85*, 331–343.

Summary

Heat transfer performance of different heat exchangers by using different nanofluids is analyzed in this literature. Effect of Reynold number, volume concentration and geometry of heat exchanger on heat exchangers is analyzed in this study. Optimum values of concentration, Reynold number and Peclet number are achieved to attain maximum heat transfer performance. From results of above literature following important points observed are:

1. Heat transfer rate was enhanced by raising flow rate, Reynold number and concentration of nanoparticles up to optimum value. Higher flow rate and Reynold number helped in enhancing Brownian motion and reducing thermal boundary layer thickness which allowed the increment in heat transfer rate.
2. Different geometric configurations of channels were applied to enhance heat exchanger efficiency. Corrugated wavy channel was applied in plate heat exchanger and HTC was enhanced to 44.7% as compared to parallel plate channel.
3. Nu number was enhanced by enhancing Reynold number, Peclet number and nanoparticle concentration because Re and Pe number enhances Brownian motion of nanoparticles and concentration enhances thermal conductivity resulted in enhancing Nusselt number. Correlation of Nusselt number was developed by applying different techniques such as curve fit, linear and non-linear techniques. Nusselt number was either dependent or independent of volume fraction as showed by relation proposed in different researches. Nusselt number was influenced by volume concentration, Re number, Pr number and other geometric parameters.
4. In double pipe heat exchanger, wire coil core rod showed better performance than simple rod. 54% better Nu number was achieved with wire coil core rod than without wire coil rod.
5. One of the main drawback of nanofluid observed was the enhancement in frictional losses and pressure drop inside the heat exchangers (tubes, pipes).

H. M. Ali et al., *Nanofluids for Heat Exchangers*,
https://doi.org/10.1007/978-981-19-3227-4

Frictional factor and pressure drop were directly related to volume concentration. Friction factor also depends on geometry of heat exchanger. For example, friction factor of twisted tape in double pipe was 80% greater than plain tube.

6. MWCNT/Solar glycol had maximum stability of 48 days while least stability was noticed for Al_2O_3–TiO_2/H_2O nanofluid as found in literature.

More research is required for use of nanofluid in heat exchangers under different geometric conditions because there is contradiction present between correlations of different research given in Table A.2. Moreover, effect of surfactant on nanofluid thermal properties should be studied because some research revealed that surfactant makes changes in thermo physical properties of nanofluid due to which results were affected. Heat transfer through nanofluid by utilizing electric and magnetic field in heat exchanging devices is attaining interest of researchers. In future, it is suggested that effect of magnetic and electric field should be studied on heat transfer applications.

Appendix

Table A.1 General correlations used for heat exchanger performance analysis using nanofluids

Refs.	Equation name	Equation
[1]	Density of nanofluid	$\rho_{\mathrm{nf}} = \varphi\rho_{\mathrm{np}} + (1 - \varphi)\rho_{\mathrm{bf}}$
[2]	Nanofluid volume concentration	$\varphi = \frac{\frac{m_{\mathrm{np}}}{\rho_{\mathrm{np}}}}{\frac{m_{\mathrm{np}}}{\rho_{\mathrm{np}}} + \frac{m_{\mathrm{bf}}}{\rho_{\mathrm{bf}}}}$
[3]	Thermal diffusivity	$\alpha_{\mathrm{nf}} = \frac{k_{\mathrm{nf}}}{\varphi C_{p\mathrm{np}}\rho_{\mathrm{np}} + (1-\varphi)\rho_{\mathrm{bf}}C_{p\mathrm{bf}}}$
[4]	Nanofluid specific heat	$C_{p\mathrm{nf}} = \frac{\varphi C_{p\mathrm{np}}\rho_{\mathrm{np}} + (1-\varphi)\rho_{\mathrm{bf}}C_{p\mathrm{bf}}}{\rho_{\mathrm{nf}}}$
[4]	Nanofluid specific heat	$C_{pnf} = \varphi C_{pnp} + (1 - \varphi)C_{pbf}$
[4]	Viscosity of nanofluid	$\mu_{\mathrm{nf}} = \left(2.1275 - 0.0215T + 0.0002T^2\right)\mu_w$
[5]	Viscosity of nanofluid	$\mu_{\mathrm{nf}} = (1 + 2.5\varphi)\mu_{\mathrm{bf}}$
[1]	Heat removed from hot liquid	$Q_h = m_h C_{p,h}(T_{hi} - T_{ho})$
[1]	Heat absorbed by cold liquid	$Q_c = m_c C_{p,c}(T_{co} - T_{ci})$
[1]	Reynold number	$\mathrm{Re} = \frac{u_s D_h \rho}{\mu};\quad u_s = \frac{V}{w b_c N} \& D_h = 2b_c$
[1]	Nusselt and Prandtl number	$\mathrm{Nu} = \frac{hD_h}{k} \& \mathrm{Pr} = \frac{C_p\mu}{k}$
[1]	Total heat transfer coefficient	$U = \frac{Q}{A \cdot F \cdot \mathrm{LMTD}}$
[1]	Log mean temperature difference	$\mathrm{LMTD} = \frac{(T_{ho} - T_{ci}) - (T_{hi} - T_{co})}{\ln\left(\frac{T_{ho} - T_{ci}}{T_{hi} - T_{co}}\right)}$
[6]	Convective heat transfer coefficient for water	$h_w = 0.295\left(\frac{k_w}{D_h}\right)\mathrm{Re}^{0.64}\,\mathrm{Pr}^{0.32}\left(\frac{\pi}{2} - \beta\right)^{0.09}$
[7]	Overall heat transfer coefficient	$\frac{1}{U_o} = \frac{1}{h_{fg}} + \frac{d_o \ln\left(\frac{d_o}{d_i}\right)}{2k_w} + \frac{1}{h_{\mathrm{nf}}}\left(\frac{d_o}{d_i}\right)$
[7]	Number of tubes outside heat transfer area	$A_s = \pi L d_o N_t$
[7]	Number of transfer units	$\mathrm{NTU} = \frac{U_o A_s}{C_{\min}};\quad C_{\min} = (mC_p)_{fg}$

(continued)

H. M. Ali et al., *Nanofluids for Heat Exchangers*,
https://doi.org/10.1007/978-981-19-3227-4

(continued)

Refs.	Equation name	Equation
[7]	Friction factor	$f = \frac{64}{\mathrm{Re_{nf}}}$forlaminar& $f = (0.790\ln(\mathrm{Re_{nf}}) - 1.64)^{-2}$forturbulent
[8]	Pressure drop	$\Delta P_h = P_{h,i} - P_{h,o}$& $\Delta P_c = P_{c,i} - P_{c,o}$
[6]	Pressure drop in ports of plate heat exchanger	$\Delta P_{\mathrm{ch}} = 4f\left(\frac{L_p}{D_h}\rho\frac{u^2}{2}\right)$
[9]	Friction factor	$f = \frac{\Delta P}{\left(\frac{L}{D_h}\right)\frac{2G^2}{\rho}}$
[8]	Friction factor proposed correlation	$f = (2.9 + 5.6\varphi + 0.12\varphi^2)\mathrm{Pe}^{-0.13}$
[10]	Nanofluid thermal conductivity	$\frac{(k_{\mathrm{eff}} - k_{\mathrm{bf}})}{k_{\mathrm{bf}}} = 3.761088\varphi + 0.017924(T - 273.15) - 0.30734$
[10]	Exergy change of hot and cold fluid	$E_h = T_e\left[C_h\ln\left(\frac{T_{h,o}}{T_{h,i}}\right)\right]$&$E_h = T_e\left[C_h\ln\left(\frac{T_{c,o}}{T_{c,i}}\right)\right]$
[11]	Exergetic effectiveness	$\varepsilon_{\mathrm{ex}} = \frac{\dot{Q}\left(1-\frac{T_o}{T_a}\right)}{Q_{\max}\left(1-\frac{T_o}{T_a}\right)_{\max}}$
[12]	Pressure drop correlation	$\Delta P = 1436.55\varphi^{0.0361}\mathrm{Re}^{0.0844}; \quad 70 \leq \mathrm{Re} \leq 220$
[2]	Overall efficiency of heat exchanger	$N = 1 - \left(\frac{A_f}{A_s}\right)(1 - n_f)$
[2]	Nanofluid thermal conductivity	$k_{\mathrm{nf}} = \left[\frac{k_{\mathrm{np}}+2k_{\mathrm{bf}}+2\varphi(k_{\mathrm{np}}-k_{\mathrm{bf}})}{k_{\mathrm{np}}+2k_{\mathrm{bf}}-\varphi(k_{\mathrm{np}}-k_{\mathrm{bf}})}\right]k_{\mathrm{bf}}$
[13]	Thermal conductivity	$\frac{k_{\mathrm{nf}}-k_{\mathrm{bf}}}{k_{\mathrm{bf}}} = (1.797 \times 10^{-2}T - 4.914)\varphi^{0.7582}$
[13]	Viscosity	$\frac{\mu_{\mathrm{nf}}-\mu_{\mathrm{bf}}}{\mu_{\mathrm{bf}}} = (3.982 \times 10^{-4}T - 4.534 \times 10^{-2})\varphi^{0.4261}$
[2]	Entropy generation	$\dot{S}_{\mathrm{gen}} = (\dot{S}_{\mathrm{gen}})_{\mathrm{HeatTransfer}} + (\dot{S}_{\mathrm{gen}})_{\mathrm{fluidfriction}}$
[14]	Entropy generation due to heat transfer	$(\dot{S}_{\mathrm{gen}})_{\mathrm{HeatTransfer}} = (mC_p)_{\mathrm{nf}}\ln\left(\frac{T_{\mathrm{nf},o}}{T_{\mathrm{nf},i}}\right) + (mC_p)_h\ln\left(\frac{T_{h,o}}{T_{h,i}}\right)$
[14]	Entropy generation due to fluid friction	$(\dot{S}_{\mathrm{gen}})_{\mathrm{fluidfriction}} = \left(\frac{\Delta P_{\mathrm{nf}}}{\rho_{\mathrm{nf}}}\right)m_{\mathrm{nf}}\frac{\left(\ln\frac{T_{\mathrm{nf},o}}{T_{\mathrm{nf},i}}\right)}{T_{\mathrm{nf},o}-T_{\mathrm{nf},i}} + \left(\frac{\Delta P_h}{\rho_h}\right)m_h\frac{\left(\ln\frac{T_{h,o}}{T_{h,i}}\right)}{T_{h,o}-T_{h,i}}$
[14]	Bejan number	$\mathrm{Be} = \frac{(\dot{S}_{\mathrm{gen}})_{\mathrm{HT}}}{(\dot{S}_{\mathrm{gen}})_{\mathrm{total}}}$

See Table A.1.

Table A.2 Correlations used and derived in literature

Refs.	Correlation	Flow	Condition
[15]	$Nu = 0.012\left(Re^{0.87} - 280\right) Pr^{0.4}$	Turbulent	
[15]	$Nu = \frac{(0.125f)(Re-1000)\,Pr}{1+12.7(0.125f)^{0.5}\left(Pr^{\frac{2}{3}}-1\right)}$	Transitional and Turbulent	$3000 < Re < 5*10^6$ $0.5 < Pr < 2000$
[16]	$Nu = 0.021Re_{nf}^{0.8}\,Pr_{nf}^{0.5}$	Turbulent	$Re = 10^4 10^5$
[17]	$Nu = 0.267Re^{0.617}\,Pr^{0.4}\left(\frac{y_o}{y}\right)^{-0.213}(1+\varphi)^{0.505}$	Turbulent	Re = 5400–15,200
[18]	$Nu = 0.0059\left(1 + 7.6286\varphi^{0.6886}Pe_d^{0.001}\right)Re_{nf}^{0.9238}\,Pr_{nf}^{0.4}$	Turbulent	Re = 10,000–25,000
[19]	$Nu = 0.074Re_{nf}^{0.707}\,Pr_{nf}^{0.385}\,\varphi^{0.074}$	Turbulent	Re = 4000–18,000
[1]	$Nu = 0.247Re^{0.66}\,Pr^{0.4}$	Turbulent	
[8]	$Nu = \left(0.26 + 0.02C' - 0.0051C'^2\right)Pe^{0.27}$	Turbulent	$Pe > 5134$
[20]	$Nu = 0.2302Re^{0.745}\,Pr^{0.4}$	Laminar	$58 < Re < 624$
[21]	$Nu = 0.3762Re^{0.6681}\,Pr^{0.4}$	Laminar	$182 < Re < 956$ $5.5 < Pr < 8$
[22]	$Nu = 0.3053Re^{0.75}\,Pr^{0.3}$	Laminar	$150 \leq Re \leq 1500$ $4 \leq Pr \leq 27$
[23]	$Nu = 0.1449Re^{0.8414}\,Pr^{0.35}\left(\frac{\mu}{\mu_w}\right)^{0.14}$		$\beta = 60°/60°$
[23]	$Nu = 0.1437Re^{0.7810}\,Pr^{0.35}\left(\frac{\mu}{\mu_w}\right)^{0.14}$		$\beta = 30°/60°$
[23]	$Nu = 0.1368Re^{0.7424}\,Pr^{0.35}\left(\frac{\mu}{\mu_w}\right)^{0.14}$		$\beta = 30°/30°$
[24]	$Nu = 0.3759Re^{0.6814}\,Pr^{n}$	Laminar	$10 < Re < 900$ $5.5 < Pr < 8.5$

(continued)

Table A.2 (continued)

Refs.	Correlation	Flow	Condition
[25]	$\mathrm{Nu} = 0.089\mathrm{De}^{0.775}\,\mathrm{Pr}^{0.4}$	Laminar	
[26]	$\mathrm{Nu} = 0.007523\mathrm{Re}^{0.8}\,\mathrm{Pr}^{0.5}(1+\varphi)^{7.6}\left(1+\frac{P}{d}\right)^{0.037}$	Turbulent	$4000 < \mathrm{Re} < 15000$ $0 < \varphi < 0.02\%$ $24.45 < \mathrm{Pr} < 32.85$ $0 < \frac{P}{d} < 0.25$
[27]	$\mathrm{Nu} = 0.00564\left(1+8.14\sqrt{\varphi}\right)\mathrm{Re}^{0.9324}\,\mathrm{Pr}^{0.38}\,\mathrm{Pe}^{0.0038}$	Laminar-Turbulent	$1000 < \mathrm{Re} < 11000$
[28]	$\mathrm{Nu} = 0.02327\mathrm{Re}^{0.8}\,\mathrm{Pr}^{0.4}(1+\varphi)^{1.969}$	Turbulent	$15000 < \mathrm{Re} < 30000$ $0 < \varphi < 0.06\%$
[28]	$\mathrm{Nu} = 0.02366\mathrm{Re}^{0.8}\,\mathrm{Pr}^{0.4}(1+\varphi)^{2.195}\left(1+\frac{p}{d}\right)^{0.09874}$; for WCCR	Turbulent	$15000 < \mathrm{Re} < 30000$ $0 < \varphi < 0.06\%$ $0 < \frac{p}{d} < 1.79$
[29]	$\mathrm{Nu} =$ $0.01783\mathrm{Re}^{0.82}\,\mathrm{Pr}^{0.46}(1+\varphi)^{2.212}(1+\mathrm{AR})^{0.0152}\left(\frac{d_h}{d_i}\right)^{-0.12}$	Turbulent	$15000 < \mathrm{Re} < 30000$ $0 < \varphi < 0.06\%$ $2.84 < \mathrm{Pr} < 3.155$ $0 < \mathrm{AR} < 4$
[30]	$\mathrm{Nu} = 0.067\mathrm{Re}^{0.71}\,\mathrm{Pr}^{0.35} + 0.0005\mathrm{Re}$	Turbulent	$5000 < \mathrm{Re} < 30000$
[31]	$\mathrm{Nu} =$ $0.03666\mathrm{Re}^{0.8204}\,\mathrm{Pr}^{0.4}\left(0.001+\frac{H}{D}\right)^{0.06281}(0.001+\varphi)^{0.04704}$	Turbulent	$10000 \le \mathrm{Re} \le 22000$ $4.4 \le \mathrm{Pr} \le 5.5$ $0 \le \varphi \le 0.5$ $0 \le H/D \le 83$
	$\mathrm{Nu} = 3.138 \times 10^{-3}(\mathrm{Re})(\mathrm{Pr})^{0.6}\left(1+\frac{H}{D}\right)^{0.03}(1+\varphi)^{1.22}$		$0 < \frac{H}{D} < 15$, $3500 < \mathrm{Re} < 8500$, $4.5 < \mathrm{Pr} < 5.5$

(continued)

Table A.2 (continued)

Refs.	Correlation	Flow	Condition
[32]	$Nu = 0.4031Re^{0.46}\,Pr^{0.33}\left(1+\frac{D}{H}\right)^{0.2781}$	Laminar	$100 < Re < 3000$ $5 < Pr < 400$ $2.5 < \frac{H}{D} < 10$
[33]	$Nu = 7.135 \times 10^{-4}Re^{0.9545}\,Pr^{0.913}\left(1+\varphi^{0.1358}\right)$	Laminar-Turbulent	$2800 < Re < 8400$ $40 < Pr < 73$
[34]	$Nu = 0.0311(Re\,Pr)^{0.68}(1+\varphi)^{95.73}$	Laminar	$Re < 2300$
[35]	$Nu = 0.00115Re^{1.050}\,Pr^{0.693}\left(1+\varphi^{0.388}\right)$	Turbulent	$2900 < Re < 8800$ $39 < Pr < 71$
[36]	$Nu = 0.2624Re^{0.5860}\,Pr^{0.3}(0.001+\varphi)^{0.07094}$	Laminar	$700 < Re < 2200$ $0 < \varphi < 0.5$
[36]	$Nu = 0.2036Re^{0.55}\,Pr^{0.3}\left(1+\frac{D}{H}\right)^{4.12}$ For Pure Liquids	Laminar	$100 < Re < 3000$ $5 < Pr < 400$ $2.5 < \frac{H}{D} < 10$
[36]	$Nu =$ $0.5652Re^{0.5004}\,Pr^{0.3}(0.001+\varphi)^{0.07060}\left(0.001+\frac{D}{H}\right)^{0.02395}$	Laminar	$700 < Re < 2200$ $0 < \varphi < 0.5$ $4.4 < Pr < 6.5$ $0\langle\frac{H}{D}\rangle 15$
[37]	$Nu = 0.085Re^{0.71}\,Pr^{0.35}$	Turbulent	
[18]	$Nu = 0.4328\left(1+11.285\varphi^{0.754}Pe^{0.218}\right)Re^{0.333}\,Pr^{0.4}$		
[38]	$Nu = \left(0.65\sqrt{De}+0.76\right)Pr^{0.175}$	Laminar	$50 < De < 2000$ $5 < Pr < 175$

(continued)

Table A.2 (continued)

Refs.	Correlation	Flow	Condition
[39]	$\mathrm{Nu} = 0.913\mathrm{De}^{0.476}\,\mathrm{Pr}^{0.20}$	Laminar	$80 < \mathrm{De} < 1200$ $0.7 < \mathrm{Pr} < 5$
[40]	$\mathrm{Nu} = 0.023\mathrm{Re}^{0.85}\,\mathrm{Pr}^{0.4}\left(\frac{d}{D_c}\right)^{0.1}$	Turbulent	$6000 < \mathrm{Re} < 65600$ $2.9 < \mathrm{Pr} < 5.7$ $\frac{D_c}{d} = 17, 104$
[41]	$\mathrm{Nu} = 4.364 + 0.283\left(\frac{x^*}{2}\right)^{-0.506}\exp\left(-\frac{41x^*}{2}\right)$	Laminar at entrance region	$x^* = \frac{2\left(\frac{x}{D}\right)}{\mathrm{Re}\cdot\mathrm{Pr}};\ x^* > 0.003$
[42]	$\mathrm{Nu} = 0.023\mathrm{Re}^{0.8}\,\mathrm{Pr}^{0.3}$	Turbulent	$\mathrm{Re} \geq 10^4$ $0.6 \leq \mathrm{Pr} \leq 160$
[43]	$\mathrm{Nu} = 0.027\mathrm{Re}^{0.8}\,\mathrm{Pr}^{\frac{1}{3}}\left(\frac{\mu}{\mu_s}\right)^{0.14}$	Turbulent	$\mathrm{Re} \geq 10^4$ $0.7 \leq \mathrm{Pr} \leq 16700$ $\frac{L}{D} \geq 10$
[44]	$\mathrm{Nu} = 0.037(\mathrm{Re}^{0.75} - 180)\,\mathrm{Pr}^{0.42}\left[1 + \left(\frac{D}{L}\right)^{\frac{2}{3}}\right]\left(\frac{\mu_b}{\mu_w}\right)^{0.14}$	Transitional and Turbulent	$2300 < \mathrm{Re} < 10^6$ $0.5 < \mathrm{Pr} < 1000$
[45]	$\mathrm{Nu} = 1.86(\mathrm{Re}\,\mathrm{Pr})^{\frac{1}{3}}\left(\frac{\mu_b}{\mu_w}\right)^{0.14}$	Laminar	
[46]	$\mathrm{Nu} = 3.66 + \frac{0.0668\left(\frac{D}{L}\right)\mathrm{Re}\,\mathrm{Pr}}{1+0.04\left[\left(\frac{D}{L}\right)\mathrm{Re}\,\mathrm{Pr}\right]^{\frac{2}{3}}}$	Laminar	
[47]	$\mathrm{Nu} = 1.953\left(\mathrm{Re}\,\mathrm{Pr}\left(\frac{D}{L}\right)\right)^{\frac{1}{3}}$	Laminar	$\mathrm{Re}\,\mathrm{Pr}\left(\frac{D}{L}\right) \geq 33.3$
[47]	$\mathrm{Nu} = 4.364 + 0.0722\left(\mathrm{Re}\,\mathrm{Pr}\left(\frac{D}{L}\right)\right)$	Laminar	$\mathrm{Re}\,\mathrm{Pr}\left(\frac{D}{L}\right) < 33.3$

(continued)

Table A.2 (continued)

Refs.	Correlation	Flow	Condition
[48]	$Nu = 5 + 0.15Re^{0.856}\,Pr^{0.347}$	Turbulent	$Re > 10000$ $0.7 \leq Pr \leq 160$
[49]	$Nu_h = 0.348Re_h^{0.663}\,Pr_h^{0.33}$		
[50]	$Nu = 0.306Re^{0.529}\,Pr^{0.33}$	Laminar	$20 < Re < 400$
[50]	$Nu = 0.562Re^{0.326}\,Pr^{0.33}$	Laminar	$Re < 20$

Nomenclature

A'C'	Amorphous Carbonic
AC	Activated Carbon
AR	Aspect Ratio
b_c	Mean Spacing Between Plates
C'	Percent Concentration
$C_{\min}$	Minimum Capacity Rate
C_p	Hot and Cold Fluid Specific heat
CMC	Carboxymethyl Cellulose
CTAB	Cetyl Trimethyl Ammonium Bromide
CWC	Corrugated Wavy Channel
Cp bf	Base fluid Specific Heat
Cp np	Nanoparticle Specific Heat
D	Diameter of Helical Coil
d	Tube side diameter
DI	Deionized
f	Friction factor
F	Temperature Correction factor
fg	Flue gas
G	Specific flow rate
GA	Gum Arabic
h	Hot fluid flow
HTC	Heat Transfer Coefficient
HTCR	Heat Transfer Coefficient Ratio
HTRR	Heat Transfer Rate Ratio
k	Thermal Conductivity
L	Length
m_{bf}	Mass of basefluid
m_{np}	Mass of nanoparticle
N	Number of passes per stream
NDG	Nitrogen doped Graphene
OHTC	Overall Heat Transfer Coefficient
OHTCR	Overall Heat Transfer Coefficient Ratio
P	Helical Pitch
Pe	Peclet number
PEC	Performance Evaluation Criteria
PVP	Polyvinyl Pyrrolidine
SDBS	Sodium dodecyl benzene sulfonate
SLS	Sodium Lauryl Sulfate
T_{a}	Thermodynamic average Temperature
TPF	Thermal Performance Factor
u_{s}	Superficial velocity
V	Volumetric flow rate
w	Plate width inside gasket

y_o/y	Overlapped Twist Ratio
β	Chevron angle of PHE
μ_b	Fluid viscosity at bulk mean temperature
μ_w	Fluid viscosity at tube wall
ρ_{bf}	Base fluid density
ρ_{np}	Nanoparticle density
φ	Volume concentration
μ_{bf}	Base fluid viscosity
μ_{nf}	Nanofluid viscosity

Bibliography

1. Pantzali, M., Mouza, A., & Paras, S. V. (2009). Investigating the efficacy of nanofluids as coolants in plate heat exchangers (PHE). *64*(14), 3290–3300.
2. Javadi, F. S., et al. (2013). The effects of nanofluid on thermophysical properties and heat transfer characteristics of a plate heat exchanger. *44*, 58–63.
3. Farajollahi, B., et al. (2010). Heat transfer of nanofluids in a shell and tube heat exchanger. *53*(1–3), 12–17.
4. Pantzali, M., et al. (2009). Effect of nanofluids on the performance of a miniature plate heat exchanger with modulated surface. *30*(4), 691–699.
5. Maré, T., et al. (2011). Comparison of the thermal performances of two nanofluids at low temperature in a plate heat exchanger. *35*(8), 1535–1543.
6. Barzegarian, R., et al. (2016). Experimental investigation on heat transfer characteristics and pressure drop of BPHE (brazed plate heat exchanger) using TiO_2–water nanofluid. *74*, 11–18.
7. Elias, M., et al. (2014). Effect of different nanoparticle shapes on shell and tube heat exchanger using different baffle angles and operated with nanofluid. *70*, 289–297.
8. Pandey, S. D., & Nema, V. K. (2012). Experimental analysis of heat transfer and friction factor of nanofluid as a coolant in a corrugated plate heat exchanger. *38*, 248–256.
9. Tiwari, A. K., Ghosh, P., & Sarkar, J. (2013). Heat transfer and pressure drop characteristics of CeO_2/water nanofluid in plate heat exchanger. *57*(1–2), 24–32.
10. Khairul, M. A., et al. (2014). Heat transfer performance and exergy analyses of a corrugated plate heat exchanger using metal oxide nanofluids. *50*, 8–14.
11. Khairul, M. A., et al. (2013). Heat transfer and thermodynamic analyses of a helically coiled heat exchanger using different types of nanofluids. *67*, 398–403.
12. Taghizadeh-Tabari, Z., et al. (2016). The study on application of TiO_2/water nanofluid in plate heat exchanger of milk pasteurization industries. *58*, 1318–1326.
13. Khoshvaght-Aliabadi, M., et al. (2014). Experimental analysis of thermal–hydraulic performance of copper–water nanofluid flow in different plate-fin channels. *52*, 248–258.
14. Kumar, V., et al. (2016). Effect of chevron angle on heat transfer performance in plate heat exchanger using ZnO/water nanofluid. *118*, 142–154.
15. Gnielinski, V. (1976). New equations for heat and mass transfer in turbulent pipe and channel flow. *16*(2), 359–368.
16. Pak, B. C., & Cho, Y. I. (1998). Hydrodynamic and heat transfer study of dispersed fluids with submicron metallic oxide particles. *11*(2), 151–170.
17. Eiamsa-Ard, S., et al. (2015). Heat transfer enhancement of TiO_2/water nanofluid in a heat exchanger tube equipped with overlapped dual twisted-tapes. *18*(3), 336–350.
18. Li, Q., & Yimin, X. (2002). Convective heat transfer and flow characteristics of Cu-water nanofluid. *45*(4), 408–416.

H. M. Ali et al., *Nanofluids for Heat Exchangers*,
https://doi.org/10.1007/978-981-19-3227-4

19. Duangthongsuk, W., & Wongwises, S. (2009). Heat transfer enhancement and pressure drop characteristics of TiO_2–water nanofluid in a double-tube counter flow heat exchanger. *52*(7–8), 2059–2067.
20. Huang, D., et al. (2015). Pressure drop and convective heat transfer of Al_2O_3/water and MWCNT/water nanofluids in a chevron plate heat exchanger. *89*, 620–626.
21. Huang, D., et al. (2016). Effects of hybrid nanofluid mixture in plate heat exchangers. *72*, 190–196.
22. Ray, D. R., et al. (2014). Experimental and numerical investigations of nanofluids performance in a compact minichannel plate heat exchanger. *71*, 732–746.
23. Elias, M., et al. (2014). Performance investigation of a plate heat exchanger using nanofluid with different chevron angle. In *Advanced Materials Research*.
24. Wang, Z., et al. (2018). Experimental comparative evaluation of a graphene nanofluid coolant in miniature plate heat exchanger. *130*, 148–156.
25. Wu, Z., Wang, L., & Sunden, B. (2013). Pressure drop and convective heat transfer of water and nanofluids in a double-pipe helical heat exchanger. *60*(1–2), 266–274.
26. Reddy, M. C. S., & Rao, V. V. (2014). Experimental investigation of heat transfer coefficient and friction factor of ethylene glycol water based TiO_2 nanofluid in double pipe heat exchanger with and without helical coil inserts. *50*, 68–76.
27. Sarafraz, M., & Hormozi, F. (2015). Intensification of forced convection heat transfer using biological nanofluid in a double-pipe heat exchanger. *66*, 279–289.
28. Sundar, L. S., et al. (2019). Heat transfer and effectiveness experimentally-based analysis of wire coil with core-rod inserted in Fe_3O_4/water nanofluid flow in a double pipe U-bend heat exchanger. *134*, 405–419.
29. Kumar, N. R., et al. (2017). Heat transfer, friction factor and effectiveness of Fe_3O_4 nanofluid flow in an inner tube of double pipe U-bend heat exchanger with and without longitudinal strip inserts. *85*, 331–343.
30. Sajadi, A., & Kazemi, M. H. (2011). Investigation of turbulent convective heat transfer and pressure drop of TiO_2/water nanofluid in circular tube. *38*(10), 1474–1478.
31. Sundar, L. S., & Sharma, K. V. (2010). Turbulent heat transfer and friction factor of Al_2O_3 nanofluid in circular tube with twisted tape inserts. *53*(7–8), 1409–1416.
32. Sarma, P., et al. (2003). Laminar convective heat transfer with twisted tape inserts in a tube. *42*(9), 821–828.
33. Hojjat, M., et al. (2011). Convective heat transfer of non-Newtonian nanofluids through a uniformly heated circular tube. *50*(4), 525–531.
34. Suresh, S., et al. (2012). Effect of Al_2O_3–Cu/water hybrid nanofluid in heat transfer. *38*, 54–60.
35. Hojjat, M., et al. (2011). Turbulent forced convection heat transfer of non-Newtonian nanofluids. *35*(7), 1351–1356.
36. Sundar, L. S., & Sharma, K. V. (2011). Laminar convective heat transfer and friction factor of Al_2O_3 nanofluid in circular tube fitted with twisted tape inserts. *3*(2), 65–278.
37. Maiga, S. E. B., et al. (2005). Heat transfer enhancement by using nanofluids in forced convection flows. *26*(4), 530–546.
38. Dravid, A. N., et al. (1971). Effect of secondary fluid motion on laminar flow heat transfer in helically coiled tubes. **17**(5), 1114–1122.
39. Kalb, C., & Seader, J. D. (1972). Heat and mass transfer phenomena for viscous flow in curved circular tubes. *15*(4), 801–817.
40. Seban, R. A., & McLaughlin, E. F. (1963). Heat transfer in tube coils with laminar and turbulent flow. *6*(5), 387–395.
41. Sharma, K., et al. (2009). Estimation of heat transfer coefficient and friction factor in the transition flow with low volume concentration of Al_2O_3 nanofluid flowing in a circular tube and with twisted tape insert. *36*(5), 503–507.
42. Manay, E., et al. (2012). Use of nanofluids in microchannels. *627*(53), 28–42.
43. Sieder, E. N., & Tate, G. E. (1936). Heat transfer and pressure drop of liquids in tubes. *28*(12), 1429–1435.

44. Hausen, H. J. A. W. (1959). Neue Gleichungen fur die Warmeubertragung bei freier oder erzwungener Stromung. *9*, 75–79.
45. Welty, J. R., et al. (2009). *Fundamentals of momentum, heat, and mass transfer*. John Wiley & Sons.
46. Nourafkan, E., Karimi, G., & Moradgholi, J. (2015). Experimental study of laminar convective heat transfer and pressure drop of cuprous oxide/water nanofluid inside a circular tube. *28*(1), 58–68.
47. Shah, R. (1975). Thermal entry length solutions for the circular tube and parallel plates. In *Third national heat mass transfer conference, Indian Institute of Technology*.
48. Notter, R., & Sleicher, C. A. (1972). A solution to the turbulent Graetz problem—III fully developed and entry region heat transfer rates. *27*(11), 2073–2093.
49. Kakaç, S., Liu, H., & Pramuanjaroenkij, A. (2002). *Heat exchangers: Selection, rating, and thermal design*. CRC Press.
50. Ayub, Z. H. (2003). Plate heat exchanger literature survey and new heat transfer and pressure drop correlations for refrigerant evaporators. *24*(5), 3–16.

The manufacturer's authorised representative in the EU is Springer Nature Customer Service Centre GmbH, Europaplatz 3, 69115 Heidelberg, Germany. If you have any concerns regarding our products, please contact ProductSafety@springernature.com

Printed and bound by CPI Group (UK) Ltd, Croydon, CR0 4YY
30/09/2026
02225259-0012